Environmental Forensics

ISSUES IN ENVIRONMENTAL SCIENCE AND TECHNOLOGY

EDITORS:

TITLES IN THE SERIES:

How to obtain future titles on publication

A subscription is available for this series. This will bring delivery of each new volume immediately on publication and also provide you with online access to each title via the Internet. For further information visit http://www.rsc.org/Publishing/Books/issues or write to the address below.

For further information please contact:
Sales and Customer Care, Royal Society of Chemistry, Thomas Graham House, Science Park, Milton Road, Cambridge, CB4 0WF, UK
Telephone: +44 (0)1223 432360, Fax: +44 (0)1223 426017, Email: sales@rsc.org

ISSUES IN ENVIRONMENTAL SCIENCE AND TECHNOLOGY

EDITORS: R.E. HESTER AND R.M. HARRISON

26
Environmental Forensics

RSCPublishing

ISBN: 978-0-85404-957-8

ISSN: 1350-7583

A catalogue record for this book is available from the British Library

Published by The Royal Society of Chemistry,
Thomas Graham House, Science Park, Milton Road,
Cambridge CB4 0WF, UK

Registered Charity Number 207890

For further information see our web site at www.rsc.org

Preface

No doubt all readers of *Issues in Environmental Science & Technology* will have a good idea of what is encompassed by forensic science and environmental science. Many, however, will have a much less clear concept of what is implied by environmental forensics. Recent decades have seen a burgeoning of legislation designed to protect the environment and, as the costs of environmental damage and clean-up are considerable, not only are there prosecutions by regulatory agencies, but also the courts are used as a means of adjudication of civil damage claims relating to environmental causes or environmental degradation. One of the editors of this volume (RMH) has been involved as an expert witness both in criminal prosecutions of companies who have breached regulations for environmental protection and in civil claims relating to harm caused by excessive pollutant releases to the environment. Such cases can become extremely protracted as expert witnesses provide their sometimes conflicting interpretations of environmental measurement data and their meaning.

In this context, it is pleasing to see that environmental forensics is developing as a specialism, leading to greater formalisation of investigative methods which should lead to more definitive findings and less scope for experts to disagree. The burgeoning of this subject is reflected in the fact that specialist degree courses in environmental forensics are available in a number of universities and a specialised journal devoted to the topic is now published.

We have been fortunate to attract some of the leading practitioners of environmental forensics to contribute chapters to this volume. The first chapter, by Stephen Mudge, gives a general introduction to the legislative framework of environmental forensics and then provides a number of case studies illustrating the applications of environmental forensics, particularly in the field of source apportionment. As this chapter makes clear, there are both chemical and biological approaches to source apportionment. The second chapter, by Andrew Ball, Jules Pretty and Rakhi Mahmud, deals with microbial techniques for environmental forensics. The explosion of techniques in molecular biology has opened up all sorts of possibilities in this area, which the chapter describes alongside the traditional microbial methods. The following chapter, by James Ehleringer and co-authors, again focuses on techniques, in this case those based upon stable isotopes. Environmental processes lead to the fractionation of isotopes of elements such as hydrogen, carbon, nitrogen, oxygen and sulfur, which can be traced with very high precision using isotope ratio mass spectrometry. The principles behind these techniques and a number of fascinating applications are outlined in this chapter. Isotope ratio methods are shown to be

Issues in Environmental Science and Technology, No. 26
Environmental Forensics
Edited by RE Hester and RM Harrison
© Royal Society of Chemistry 2008

effective in tracing the sources both of wine and cocaine. In the fourth chapter, Scott Stout and Zhendi Wang describe the chemical composition of petroleum and petroleum products and show how diagnostic compounds can be used for fingerprinting petroleum in the environment. The complex chemical make-up of petroleum, which may contain tens of thousands of individual hydrocarbons and other compounds, is shown to provide many opportunities for chemical fingerprinting, which has many applications in identifying the sources of petroleum products in the environment.

In the fifth chapter, Ioana Petrisor and Jim Wells address the question of whether perchlorate in the environment is natural or man-made. The chapter provides a detailed forensic guide to identification of the sources of this important pollutant. In the following chapter, the same authors deal with the tracking of chlorinated solvents in the environment. These compounds have been very widely used and are the most frequently detected groundwater contaminants in the USA and most probably in many other countries. Forensic methods, which can take account of differential rates of decay in the environment, provide an essential means of tracking and source identification for chlorinated solvents. In the final chapter, Stanley Feenstra and Michael Rivett explore the emerging role of environmental forensics in the investigation of groundwater pollution. They deal with a wide range of techniques applicable to key pollutants of groundwater for which source identification is necessary either to identify those responsible for the pollution and remediation or for the purpose of closing off an undesirable discharge.

This volume provides a comprehensive overview of many of the key areas of environmental forensics written by some of the leading practitioners in the field. It will prove to be both of specialist use to those seeking expert insights into the field and its capabilities and also of more general interest to those involved in both environmental analytical science and environmental law.

Ronald E. Hester
Roy M. Harrison

Contents

Issues in Environmental Science and Technology, No. 26
Environmental Forensics
Edited by RE Hester and RM Harrison
© Royal Society of Chemistry 2008

Spatial Considerations of Stable Isotope Analyses in Environmental Forensics
James R. Ehleringer, Thure E. Cerling, Jason B. West,
David W. Podlesak, Lesley A. Chesson and Gabriel J. Bowen

Diagnostic Compounds for Fingerprinting Petroleum in the Environment
Scott A. Stout and Zhendi Wang

Perchlorate – Is Nature the Main Manufacturer?
Ioana G. Petrisor and James T. Wells

Tracking Chlorinated Solvents in the Environment
Ioana G. Petrisor and James T. Wells

Groundwater Pollution: The Emerging Role of Environmental Forensics
Stanley Feenstra and Michael O. Rivett

Editors

Ronald E. Hester, BSc, DSc(London), PhD(Cornell), FRSC, CChem

Ronald E. Hester is now Emeritus Professor of Chemistry in the University of York. He was for short periods a research fellow in Cambridge and an assistant professor at Cornell before being appointed to a lectureship in chemistry in York in 1965. He was a full professor in York from 1983 to 2001. His more than 300 publications are mainly in the area of vibrational spectroscopy, latterly focusing on time-resolved studies of photoreaction intermediates and on biomolecular systems in solution. He is active in environmental chemistry and is a founder member and former chairman of the Environment Group of the Royal Society of Chemistry and editor of 'Industry and the Environment in Perspective' (RSC, 1983) and 'Understanding Our Environment' (RSC, 1986). As a member of the Council of the UK Science and Engineering Research Council and several of its sub-committees, panels and boards, he has been heavily involved in national science policy and administration. He was, from 1991 to 1993, a member of the UK Department of the Environment Advisory Committee on Hazardous Substances and from 1995 to 2000 was a member of the Publications and Information Board of the Royal Society of Chemistry.

Roy M. Harrison, BSc, PhD, DSc(Birmingham), FRSC, CChem, FRMetS, Hon MFPH, Hon FFOM

Roy M. Harrison is Queen Elizabeth II Birmingham Centenary Professor of Environmental Health in the University of Birmingham. He was previously Lecturer in Environmental Sciences at the University of Lancaster and Reader and Director of the Institute of Aerosol Science at the University of Essex. His more than 300 publications are mainly in the field of environmental chemistry, although his current work includes studies of human health impacts of atmospheric pollutants as well as research into the chemistry of pollution phenomena. He is a past Chairman of the Environment Group of the Royal Society of Chemistry for whom he has edited 'Pollution: Causes, Effects and Control' (RSC, 1983; Fourth Edition, 2001) and 'Understanding our Environment: An Introduction to Environmental Chemistry and Pollution' (RSC, Third Edition, 1999). He has a close interest in scientific and policy aspects of air pollution, having been Chairman of the Department of Environment Quality of Urban Air Review Group and the DETR Atmospheric Particles Expert Group. He is currently a member of the DEFRA Air Quality Expert Group, the DEFRA Expert Panel on Air Quality Standards, and the Department of Health Committee on the Medical Effects of Air Pollutants.

List of Contributors

Andrew S. Ball, School of Biological Sciences, Flinders University, Sturt Road, Bedford Park, Adelaide, SA 5001, Australia

Gabriel J. Bowen, Purdue University, West Lafayette, IN 47907, USA

Thure E. Cerling, IsoForensics Inc., 423 Wakara Way, Salt Lake City, UT 84108 and University of Utah, Salt Lake City, UT 84112, USA

Lesley A. Chesson, IsoForensics Inc., 423 Wakara Way, Salt Lake City, UT 84108 and University of Utah, Salt Lake City, UT 84112, USA

James R. Ehleringer, IsoForensics Inc., 423 Wakara Way, Salt Lake City, UT 84108 and University of Utah, Salt Lake City, UT 84112, USA

Stanley Feenstra, Applied Groundwater Research Ltd., 5285 Drenkelly Court, Mississauga, Ontario, Canada L5M 2H7

Rakhi Mahmud, Department of Biological Sciences, University of Essex, Wivenhoe Park, Colchester CO4 3SQ, UK

Stephen M. Mudge, Marine Chemistry and Environmental Forensics, School of Ocean Sciences, University of Wales – Bangor, Menai Bridge, Anglesey LL59 5AB, UK

Ioana G. Petrisor, Haley & Aldrich, 3187 Red Hill Ave., Suite 155, Costa Mesa, CA 92626, USA

David W. Podlesak, IsoForensics Inc., 423 Wakara Way, Salt Lake City, UT 84108 and University of Utah, Salt Lake City, UT 84112, USA

Jules N. Pretty, Department of Biological Sciences, University of Essex, Wivenhoe Park, Colchester CO4 3SQ, UK

Michael O. Rivett, School of Geography, Earth and Environmental Sciences, University of Birmingham, Birmingham B15 2TT, UK

Scott A. Stout, NewFields Environmental Forensics Practice LLC, 300 Ledgewood Place, Suite 305, Rockland, MA 02370, USA

Zhendi Wang, Oil Spill Research, ESTD, Environmental Technology Center, Environmental Canada, 335 River Road, Ottawa, Ontario, Canada K1A 0H3

James T. Wells, Haley & Aldrich, 3700 State Street, Suite 350, Santa Barbara, CA 93105, USA

Jason B. West, University of Utah, Salt Lake City, UT 84112, USA

Environmental Forensics and the Importance of Source Identification

STEPHEN M. MUDGE

1 Introduction

The scientific community has been taking samples and identifying the source of the materials in those samples for many years – this is the basis of all environmental forensics, the identification and source apportionment of compounds in environmental samples. What is different, however, is that with source apportionments comes blame, and blame these days means costs.

Environmental forensics could be summarized as an investigation of what is in the environment, where it has come from and using that data to prosecute those who have contravened particular laws. There are other aspects to environmental forensics which include data mining and prediction to understand better what is going on, helping industries at the design stage to ensure they comply with relevant legislation and simply reconstructing environmental histories – who did what, when and where?

2 The Legislative Framework for Environmental Forensics

One of the key aspects of environmental forensics is the bringing together of data in order to have a successful prosecution. There are many debates as to what prosecutions are for (deterrent, punishment, revenge, *etc.*), but they are beyond the scope of this chapter; however, there is a considerable framework in law to enable either the State or individuals to seek redress with regard to contamination or injury due to damage to the environment.[1]

2.1 National Legislation

2.1.1 UK. Most of the recent legislation protecting the environment in the UK has been derived from European Directives, which are discussed later. However, prior to 1995, there were several key pieces of legislation that have

Issues in Environmental Science and Technology, No. 26
Environmental Forensics
Edited by RE Hester and RM Harrison
© Royal Society of Chemistry 2008

formed the basis of many prosecutions by the State through the Environment Agency or the National Rivers Authority, its precursor. The Environment Act 1995 brought the United Kingdom Environment Agency and Scottish Environment Protection Agency (SEPA) into being and outlined the requirements of a range of bodies with regard to contaminated land and abandoned mines to enhance some aspects of pollution control.

The majority of actions against environmental crimes in the UK appear to be under this Act or the Water Resources Act 1991. In this Act, there are several sections concerned with management of water resources, but a key section with regard to environmental forensics is Section 85, Offences of Polluting Controlled Waters. The subsections of this Act are the most frequently cited offences in environmental cases (N. Evans, personal communication).

(1) A person contravenes this section if he causes or knowingly permits any poisonous, noxious or polluting matter or any solid waste matter to enter any controlled waters.

(2) A person contravenes this section if he causes or knowingly permits any matter, other than trade effluent or sewage effluent, to enter controlled waters by being discharged from a drain or sewer in contravention of a prohibition imposed under section 86 below.

(3) A person contravenes this section if he causes or knowingly permits any trade effluent or sewage effluent to be discharged –
 (a) into any controlled waters; or
 (b) from land in England and Wales, through a pipe, into the sea outside the seaward limits of controlled waters.

(4) A person contravenes this section if he causes or knowingly permits any trade effluent or sewage effluent to be discharged, in contravention of any prohibition imposed under section 86 below, from a building or from any fixed plant –
 (a) on to or into any land; or
 (b) into any waters of a lake or pond which are not inland freshwaters.

(5) A person contravenes this section if he causes or knowingly permits any matter whatever to enter any inland freshwaters so as to tend (either directly or in combination with other matter which he or another person causes or permits to enter those waters) to impede the proper flow of the waters in a manner leading or likely to lead, to a substantial aggravation of –
 (a) pollution due to other causes; or
 (b) the consequences of such pollution.

(6) Subject to the following provisions of this Chapter, a person who contravenes this section or the conditions of any consent given under this Chapter for the purposes of this section shall be guilty of an offence and liable –
 (a) on summary conviction, to imprisonment for a term not exceeding three months or to a fine not exceeding £20 000 or to both;
 (b) on conviction on indictment, to imprisonment for a term not exceeding two years or to a fine or to both.

Full copies of the UK Acts can be found at http://www.opsi.gov.uk.

Section 85(1) of the Water Resources Act 1991 provides the over-arching legislative power to bring a criminal prosecution against individuals or companies if they allow 'poisonous, noxious or polluting matter' to enter controlled waters. To bring such a prosecution, however, it would be necessary to demonstrate 'beyond all reasonable doubt' that a discharge had been made by a person or company to controlled waters. Notwithstanding the considerable amount of case law that has developed regarding the meaning of individual words in this section of the Act,[1] the science required to bring a successful case must unambiguously identify the source of the contamination and also make estimates of the amount discharged to demonstrate that it exceeded an authorization to discharge.

There are other UK Acts that are used in environmental prosecutions, including the Water Industry Act 1991. Much of this Act is to do with the supply of water and provision of sewerage services, although Section 111 sets out the restrictions on the use of public sewers and specifically forbids the discharge to public sewers of items that may lead to failure of the treatment processes or lead to unauthorized discharges from the sewage treatment works.

---

(1) Subject to the provisions of Chapter III of this Part, no person shall throw, empty or turn or suffer or permit to be thrown or emptied or to pass, into any public sewer or into any drain or sewer communicating with a public sewer –

  (a) any matter likely to injure the sewer or drain, to interfere with the free flow of its contents or to affect prejudicially the treatment and disposal of its contents; or

  (b) any such chemical refuse or waste steam or any such liquid of a temperature higher than one hundred and ten degrees Fahrenheit, as by virtue of subsection (2) below is a prohibited substance; or

  (c) any petroleum spirit or carbide of calcium.

(2) For the purposes of subsection (1) above, chemical refuse, waste steam or a liquid of a temperature higher than that mentioned in that subsection is a prohibited substance if (either alone or in combination with the contents of the sewer or drain in question) it is or, in the case of the liquid, is when so heated –

  (a) dangerous;

  (b) the cause of a nuisance; or

  (c) injurious, or likely to cause injury, to health.

(3) A person who contravenes any of the provisions of this section shall be guilty of an offence and liable –

  (a) on summary conviction, to a fine not exceeding the statutory maximum and to a further fine not exceeding £50 for each day on which the offence continues after conviction;

  (b) on conviction on indictment, to imprisonment for a term not exceeding two years or to a fine or to both.

(4) For the purposes of so much of subsection (3) above as makes provision for the imposition of a daily penalty –
  (a) the court by which a person is convicted of the original offence may fix a reasonable date from the date of conviction for compliance by the defendant with any directions given by the court; and
  (b) where a court has fixed such a period, the daily penalty shall not be imposed in respect of any day before the end of that period.
(5) In this section the expression "petroleum spirit" means any such –
  (a) crude petroleum;
  (b) oil made from petroleum or from coal, shale, peat or other bituminous substances; or
  (c) product of petroleum or mixture containing petroleum as, when tested in the manner prescribed by or under the [1928 c. 32.] Petroleum (Consolidation) Act 1928, gives off an inflammable vapour at a temperature of less than seventy-three degrees Fahrenheit.

This Act provides the mechanism by which Water Companies may pursue people who discharge unauthorized materials into sewers, the principal method by which waste is often removed from industrial sites. Subsection (1) sets out the general condition and the other subsections refine that clause further.

2.2   Regional Legislation

2.2.1   European Union Directives.   In recent years, much of the environmental legislation enacted in the UK and across other European countries has been derived from EU Directives. Directives require Member States of the EU to implement legislation to achieve a particular outcome but do not specify how the law should be written. This gives the individual countries the opportunity to implement the Directive in the manner that best suits them while achieving the intended outcomes. Key environmental Directives that have been adopted or in the process of being transposed into national law include:

- (Amended) Bathing Waters (2006/7/EC)
- Shellfish Waters (79/923/EEC)
- Water Framework Directive (2000/60/EC)
- Habitats (92/43/EEC)
- Waste Incineration (2000/76/EC)
- Environmental Liability (2004/35/CE)
- Air Quality Framework Directive (96/62/EC).

Although this list is not exhaustive, it does highlight the role that these transnational Directives have in improving the legislative framework by which the environment across Europe is protected. At the heart of most of these Directives is the Precautionary Principle, the idea that in the absence of scientific consensus, caution should be applied and discharges or activities curtailed or the burden for subsequent effects is accepted by the person wishing to discharge.

The Environmental Liability Directive is new piece of legislation which promises to enhance the mechanisms by which 'the polluter pays' principle is enhanced. In summary, the Directive is aimed at the prevention and remedying of environmental damage – specifically, damage to habitats and species protected by EC law, damage to water resources and land contamination which presents a threat to human health. It would apply only to damage from incidents occurring after it comes into force.[2]

Important aspects include:[2]

1. It is based on 'the polluter pays' principle, *i.e.* polluters should bear the cost of remediating the damage they cause to the environment or of measures to prevent imminent threat of damage.
2. Polluters would meet their liability by remediating the damaged environment directly or by taking measures to prevent imminent damage or by reimbursing competent authorities who, in default, remediate the damage or take action to prevent damage.
3. Competent authorities would be responsible for enforcing the regime in the public interest, including determining remediation standards or taking action to remediate or prevent damage and recover the costs from the operator.
4. Strict liability would apply in respect of damage to land, water and biodiversity from activities regulated by specified EU legislation; fault-based liability would apply in respect of biodiversity damage from any other activity.
5. Defences would exist for damage caused by an act of armed conflict, natural phenomenon or from compliance with a permit and emissions which at the time they were authorized were not considered to be harmful according to the best available scientific and technical knowledge.
6. Where an operator is not liable, the Member State would have subsidiary responsibility for remediating that damage.
7. Individuals and others who may be directly affected by actual or possible damage and qualified entities (non-governmental organizations) may request action by a competent authority and seek judicial review of the authority's action or inaction.

Aspects of this legislation fit well with environmental forensic investigations: identifying the source of any contamination and demonstrating harm has been caused or may be caused at a future date. It also provides for State responsibility in a manner similar to that of the USA's CERCLA (The Comprehensive Environmental Response, Compensation and Liability Act), also known as Superfund.

2.3   US Legislation

CERCLA[3] is funded through a tax on the chemical and oil industries in the USA and the income is used to remediate abandoned sites or sites where responsibility cannot be enforced. Where responsibility for contamination can be identified, it empowers the State to use the Environmental Protection Agency (EPA) to clean up and seek financial redress from these parties. This process has created considerable environmental forensic investigations as tens of thousands of sites have been assessed and cleaned up as appropriate.

The European approach to discharge is somewhat different to one piece of key US legislation, the Clean Water Act (2002 amended). Section 303 includes the following control mechanism for contaminant discharge:

> (C) Each State shall establish for the waters identified in paragraph (1)(A) of this subsection and in accordance with the priority ranking, the total maximum daily load, for those pollutants which the Administrator identifies under section 304(a)(2) as suitable for such calculation. Such load shall be established at a level necessary to implement the applicable water quality standards with seasonal variations and a margin of safety which takes into account any lack of knowledge concerning the relationship between effluent limitations and water quality.

States are required to establish a total maximum daily load (essentially an assimilative capacity) although a caveat regarding lack of knowledge and safety margin is also added. This allows for States to determine on the basis of the local physico-chemical conditions how much of any one particular contaminant could safely be disposed of through discharges to waters.

3 Source Identification

As can be seen in the legislative frameworks outlined above, being able to identify correctly the parties (individual or companies) responsible for contamination is fundamental to success. Part of the remit of any practitioner of environmental forensics is to identify the origin of contaminants, which may be chemical or biological in nature, and demonstrate a pathway by which those materials may have reached that location. This is not as easy a task as it first appears due to the diverse nature of the chemicals and processes that take place in the environment. The ideal scheme is shown as case 1 in Figure 1 together with some complications (cases 2–5) to that scheme.

In the first case, with a single source of the contaminant and a single process or route by which that material gets to the receptor in the environment, the job of identifying the components of the system is relatively easy. In such cases, simple presence or concentration information maybe sufficient to demonstrate where the material has come from. Cases such as this are often restricted to man-made chemicals with very few possible sources and little or no material pre-existing in the environment.

However, in the second case, with multiple potential or actual sources of the contaminant, distinguishing one source from another adds to the complexity. This scheme is further complicated by the possibility of natural or historical

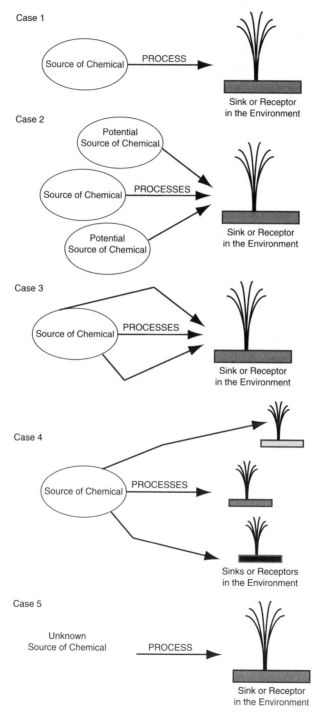

Figure 1   Scenarios for the source → process → receptor system with which environmental forensic practitioners may be involved.

occurrence of the chemical in the environment independent of any contamination event. There is a wide range of methods available to assist in distinguishing between sources and includes complex chemical signatures,[4] DNA associated with biota,[5] stable isotopes,[6] *etc.* If the potential sources produce different mixtures of materials, simple ratios between key compounds may be suitable to distinguish between each. If the sources were all of a similar nature, *e.g.* sewage discharges, complex signature analysis relying on relatively subtle differences would be needed to differentiate unambiguously between potentially responsible parties.

In the third case (Figure 1), one source of contamination may have several different routes by which the compounds reach the receptor sites. These may be of different path lengths or even through different media (air *versus* water) and if the chemical or chemicals are degradable in any way, the signature may change with time and path. It is also possible that there is a history of contamination at a site with different sources responsible at different times. Therefore, it is important to have a handle that tells us something about the age of the discharge. Differential degradation rates of components in a mixed chemical source (*e.g.* BTEX components in petroleum-based fuels) may be able to provide some information on the time since discharge, but environmental processes may differ between locations and it may be that only relative values may be obtainable.[7]

In case 4, one source may contribute to several different receptors and the concentration and signature at each may be determined by the mechanism and time of the process that carries the compounds to the sinks. Analysis of the suite of compounds at site 1 may indicate a different chemical signature compared with site 2 or 3; this difference, however, is an artefact or an indicator of the environmental transport process. A good example of this type of case is an oil spillage. Longer residence times in the environment will lead to the evaporation of the volatile components and simple GC–FID analysis may indicate two completely different aliphatic hydrocarbon profiles.[8] Careful choice of the key analytes is required in such situations and they must be conservative if being used to indicate source but have differential degradation if being used to indicate time in the environment, as with the BTEX components in case 3.

Many environmental forensic investigations fall in to case 5: a contaminant is present at a site but it is not at all clear where that material has come from. There are several sub-cases that fall into this category.

3.1   Illegal Discharges

Some unscrupulous companies or individuals may choose to dispose of their wastes in an uncontrolled or illegal manner, leading to contamination of the environment. These activities are often done in secret to avoid detection by the authorities and this may make their characterization relatively hard. In some cases, it is not apparent until significantly later after such disposals have taken place. In some cases, this may run into decades.

3.2　Fugitive Emissions or Discharge

In contrast to the above situation, these emissions may be taking place without the knowledge of the owner of the source. Examples of this type of spread of contamination include wind blow of on-site dusts and the wheels of cars and lorries picking up compounds on-site and then transferring them off-site. There is no malicious intent in this case, although these materials are making it off-site and the company is still liable under 'Strict Liability'.[1] One on-going problem is leaks from underground storage tanks such as those containing fuels.[9,10] Corrosion or physical damage may lead to punctures and the loss of the fuel, albeit at a slow rate, to soil. Differential solubility between the aliphatic components and additives such as MTBE lead to a separation of chemicals and the more water-soluble compounds may emerge at one location while the others may be retained or exit from the soil by a different route such as evaporation.

3.3　Deliberate 'Fly-tipping'

Fly-tipping is the disposal of waste at an off-site location usually in one-off events and is often associated with small companies or individuals seeking to avoid the costs of disposing at correctly licensed sites. In the current climate, such wastes often include asbestos. This type of activity may occur at a range of locations but typically they occur close to roads where easy access is present.

3.4　Historical Discharges

Many countries have a significant industrial heritage and, since the industrial revolution began in ∼1750, there may have been several industries of different types occupying the same parcel of land. Previous practices may not have been as good as now and losses to the atmosphere (dusts and smoke), waters and land may all have become contaminated with chemicals that have appreciable environmental half-lives. That means they are still around today and re-development of such land may reveal suites of chemicals from this industrial past. Apportioning responsibility for this contamination varies depending on which country it is in and whether the companies responsible are still trading after potentially 100 years.

3.5　Altered Environmental Processes

Historical contamination may be 'locked-up' in the sediments and soils out of our currently accessible regions. Natural or anthropogenically induced changes in environmental processes such as changes to the wave or current regime may erode sediments laid down decades ago and bring back into the current environmentally accessible region chemicals we have not experienced in the recent past. Typical examples include erosion of old salt marshes which were laid down when the concentration of contaminants were much greater.[11] Erosion of these fine-grained sediments may increase the concentration of the surface sediments with contaminants such as mercury and radionuclides.

4   Tools for Source Apportionment

A considerable diversity of techniques have been developed to assist in the identification of source or origin. Many of these are chemical in nature, but a few use biological attributes such as changes in the biotic community[12–14] or molecular DNA methods.[15] Simple methods may work only in simple cases and more complicated cases often require a whole range of different analyses in order to provide evidence that passes the test of 'beyond all reasonable doubt'. In these cases, a multi-evidential approach is most likely to succeed.[16]

4.1   Chemical Approaches

Most environmental investigations centre on chemical analysis of a range of different media and establishing the origin of the compounds within the matrix. It must be appreciated that the majority of magistrates and judges have not had an extensive environmental chemistry training and simple, convincing, analyses are therefore the most scientifically and legally defensible.

4.1.1   Presence or Absence.
Some chemicals are only produced synthetically and do not occur in nature. In this case, the simple presence of this compound at any location implies that it has been released from an anthropogenic source. Tracking this back to its source may be easy if there is a concentration gradient to follow back to the discharge site, but several conditions may conspire to make that more difficult than it seems. Water solubility is a key characteristic of chemicals that affect their mobility; hydrophobic compounds will tend to associate with the solid phase or partition into biota and will not necessarily move through the environment at the same rate as water flow. Conversely, hydrophilic compounds such as MTBE will readily mix with water and migrate down the hydraulic gradient.[17] A study in Portugal[4] demonstrated that the principal deposition site for sewage discharged into a lagoon was several kilometres distant from the source and no gradient for the lipophilic compounds could be discerned.

4.1.2   Ratios.
The differential solubility and volatility of components within a source mixture will lead to separation down the spill axis.[18] Volatile components will be lost to the atmosphere, leaving larger compounds behind; water-soluble components will move with the water, leaving the less water-soluble components bound to sediments or soils; degradable compounds will be metabolized into other compounds or become part of the biomass of the microbial community, leaving the more refractile compounds in the system. However, the latter process can be beneficial when used as a mechanism to aid in the dating of spills.[19] A good example of this is age dating of gasoline spills using the differential loss rates of components of the BTEX group (benzene, toluene, ethylbenzene and xylene). Due to relatively rapid losses of the parent and monomethyl derivative compared with the ethyl and trimethyl components, a ratio can be used to suggest an age since release (Figure 2).

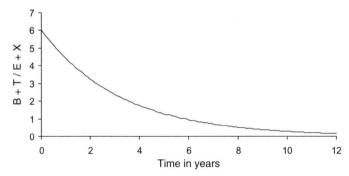

Figure 2   Relationship with time between the smaller benzene and toluene (B + T) components and ethylbenzene and xylene (E + X) components in gasoline spills (redrawn from Kaplan[19]).

There are a number of issues related to the use of this particular approach, such as initial BT/EX ratios in the fuel and different environmental conditions, but it illustrates the applicability of differential degradation rates in age dating contamination events.

4.1.3   Complex Signatures.   Sometimes there are several potential sources ('potentially responsible parties' in the jargon of CERCLA) and there may be subtle differences between the complex signatures in each. These differences can be used to differentiate between each and post-analytical statistical methods such as partial least squares (PLS) can apportion numerical proportions associated with them.[20,21] The most frequently used multivariate method for such differentiation is the projection method of principal components analysis (PCA), and this may demonstrate simply and clearly in a two-dimensional axis scheme the separation of potential sources and the environmentally measured suites of chemicals.[22] There are many hazards associated with the use of these methods[23] related to the type of data and distribution, abundance differences, *etc.*, but these can all be overcome and allow the presentation of scientifically defensible results to a court. Other related methods include factor analysis[24] and cluster analysis and multi-dimensional scaling.[25]

The use of statistics in a court situation is commonplace with DNA analysis and linking materials found at a crime scene to an individual, but when it comes to comparisons and more complex analysis, there is a danger that they will (a) not be applied properly, (b) not be understood correctly by the lay members of the court and (c) be used to obfuscate and hide the truth. The rationale for using the statistical approach must be set out clearly and the current 'best-practice' used. Clear, concise figures built from defensible analyses are usually the most effective in convincing people.

4.1.4   Change of State.   In some cases, clear gradients are not seen due to a combination of physical and chemical effects: in the case of sewage-derived deposits in the Ria Formosa Lagoon, Portugal, low water solubility of the key sterol/stanol markers meant that these compounds were carried in suspension and deposited at sites where the current velocity was low.[4] In some environments, however, the natural changes that occur directly influence the water solubility of compounds. A good example of this is triclosan [5-chloro-2-(2,4-dichlorophenoxy)phenol], an antibacterial agent used in many personal hygiene products such as toothpaste.[26] This particular compound is an example of a hydrophobic ionogenic compound that may be ionized by the environment and can be found in more that one chemical species[27] potentially with different solubilities. Some herbicides, pesticides and drugs fall in to this category and as they pass from essentially freshwater environments through a sewage treatment plant to estuarine receiving waters, the compounds may become immobilized on the sediments. However, with a change in the tide, they may become mobile again.

When tracking sources, water solubility is a key factor as water is the usual vector that moves compounds around the environment (ground waters, rivers, *etc.*). One compound that has been used in the tracking of human sewage discharges is caffeine, as it may be safely assumed that humans are the only species to drink the compound in an appreciable quantity. Caffeine is water soluble ($\sim 22 \, \mathrm{g \, l^{-1}}$) and will move with the water, but if ratios are developed with other compounds (*e.g.* stanol) that are not water soluble, the results become meaningless at best and deceptive at worst.[28]

4.1.5 Stable Isotopes. In a few cases, it may not be possible to determine unambiguously the source of the compound(s) based on their presence alone or even on their association in particular mixtures. In these cases, a more fundamental property of the compound is needed. Stable isotopes can provide that specificity in some instances.[29] Stable isotope analyses are usually carried out for the carbon atom and reported as $\delta^{13}C$ values, although several other elements may be investigated (H, S, Cl, O, N and Br). The technique relies on a difference in the relative proportion of heavy and light isotopes in the compounds from different sources such that they may be distinguished from each other.[30] These methods are routinely used in environmental forensic cases[31] and may become more useful with wider deployment of compound-specific multi-element equipment. Two-dimensional analyses (*e.g.* carbon and hydrogen) can increase the specificity of the approach[17] and will have a greater resolving power and, therefore, a better likelihood of success in a court case.

4.2 Biological Approaches

In some cases, chemical analyses are either not practical or not possible. There are instances when the chemical discharge took place in the distant past and has either been metabolized (nutrients, milk, *etc.*) or has simply been dispersed with water flow to concentrations below limits of detection (*e.g.* caffeine). If the spill had been known about at the time, interceptors could have been installed or samples could have been taken. However, the biotic assemblage living in the environment where the spill occurred will have responded to the spillage/discharge and may be altered in comparison with an assemblage unaffected by the spill.

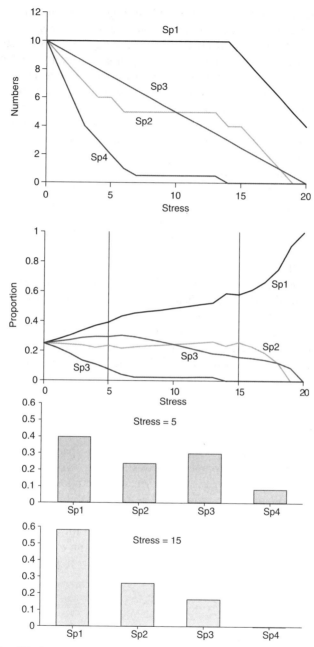

Figure 3 Simplified community system where environmental changes (stress) lead to fewer organisms. Species 1 is the least sensitive to the stress whereas species 4 is most sensitive. When converted to proportions, a different profile is evident and if samples were taken at 'Stress 5' and 'Stress 15', the assemblage will be different and may be indicative of the stressor.

This approach can be seen simplistically in Figure 3. In this case, four species live in an area; as they are stressed by a chemical discharge, for example, the species respond differently and selectively die – this may require a substantial period depending on many factors. In general, small organisms with a relatively rapid generation time are best for this approach. In marine sediments, the meiofauna proved to be very useful.[13] Indeed, in this particular example, the community structure was determined at known sewage discharge points; when this 'signature' was applied to the communities from a wide area, previously unrecognized sewage discharges were identified and relative contributions could be assessed.

Different stressors and different stressor concentrations lead to a change in the assemblage. Some organisms will be inherently more adapted or resilient to particular contaminants whereas others will 'curl up and die'. For example, at stress level 5, the profile of the organisms is different to that at stress level 15. However, the Harvard Law of Animal Behaviour ('under carefully controlled experimental circumstances, an animal will behave as it damned well pleases') must be considered and natural variability, especially genetic diversity, will lead to a spread of responses.

This approach has been developed for European marine soft sediments by Borja and co-workers[32,33] and is being used in the implementation of the Water Framework Directive.

Bacteria and algae respond to contamination in the environment in a similar way to meio- and macrofauna. The soil bacterial assemblage can vary according to the nutrient status of the soil[34] and may be used diagnostically in determining the past environmental history of that ground.

5   Summary

Environmental forensics as a subject is developing apace, with new legal instruments and regulations to spur on investigations. New tools are becoming available which are providing ever more sensitive methods to identify chemicals or changes in biological communities. Key to successful participation in expert witness cases involving the environment is good, scientifically defensible data from a wide range of approaches which all point to the same outcome. These data may need to be presented to lay members of a jury or magistrates bench, so clear, concise figures are usually best. However, if you cannot defend it scientifically, it is of no benefit to anyone.

References

1. B. Jones and N. Parpworth, *Environmental Liabilities*, Shaw, Crayford, 2004.
2. DEFRA, Environmental liability, http://www.defra.gov.uk/environment/ liability/index.htm, accessed 18 October 2007.
3. USEPA, Superfund (CERCLA), http://www.epa.gov/superfund/index.htm, accessed 18 October 2007.
4. S. M. Mudge and C. E. Duce, Identifying the source, transport path and sinks of sewage derived organic matter, *Environ. Pollut.*, 2005, **136**, 209–220.

5. D. A. Bossio, M. S. Girvan, L. Verchot, J. Bullimore, T. Borelli, A. Albrecht, K. M. Scow, A. S. Ball, J. N. Pretty and A. M. Osborn, Soil microbial community response to land use change in an agricultural landscape of Western Kenya, *Microb. Ecol.*, 2005, **49**, 50–62.

6. R. Kolhatkar, T. Kuder, P. Philp, J. Allen and J. T. Wilson, Use of compound-specific stable carbon isotope analyses to demonstrate anaerobic biodegradation of MTBE in groundwater at a gasoline release site, *Environ. Sci. Technol.*, 2002, **36**, 5139–5146.

7. Z. D. Wang, K. Li, M. Fingas, L. Sigouin and L. Menard, Characterization and source identification of hydrocarbons in water samples using multiple analytical techniques, *J. Chromatogr. A*, 2002, **971**, 173–184.

8. Z. D. Wang, S. A. Stout and M. Fingas, Forensic fingerprinting of biomarkers for oil spill characterization and source identification, *Environ. Forensics*, 2006, **7**, 105–146.

9. T. O. McGarity, MTBE: a precuationary tale, *Harvard Environ. Law Rev.*, 2004, **28**, 281–342.

10. Y. Zhang, I. A. Khan, X. H. Chen and R. F. Spalding, Transport and degradation of ethanol in groundwater, *J. Contam. Hydrol.*, 2006, **82**, 183–194.

11. B. J. Harland, D. Taylor and K. Wither, The distribution of mercury and other trace metals in the sediments of the Mersey Estuary over 25 years 1974–1998, *Sci. Total Environ.*, 2000, **253**, 45–62.

12. A. Borja, I. Muxika and J. Franco, The application of a Marine Biotic Index to different impact sources affecting soft-bottom benthic communities along European coasts, *Mar. Pollut. Bull.*, 2003, **46**, 835–845.

13. E. J. Hewitt and S. M. Mudge, Detecting anthropogenic stress in an ecosystem: 1. Meiofauna in a sewage gradient, *Environ. Forensics*, 2004, **5**, 155–170.

14. F. E. Hopkins and S. M. Mudge, Detecting anthropogenic stress in an ecosystem: 2. Macrofauna in a sewage gradient, *Environ. Forensics*, 2004, **5**, 213–223.

15. M. S. Girvan, J. Bullimore, J. N. Pretty, A. M. Osborn and A. S. Ball, Soil type is the primary determinant of the composition of the total and active bacterial communities in arable soils, *Appl. Environ. Microbiol.*, 2003, **69**, 1800–1809.

16. C. Allen, *Practical Guide to Evidence*, Cavendish Publishing, London, 2001.

17. A. Davis, B. Howe, A. Nicholson, S. McCaffery and K. A. Hoenke, Use of geochemical forensics to determine release eras of petrochemicals to groundwater, Whitehorse, Yukon, *Environ. Forensics*, 2005, **6**, 253–271.

18. G. S. Douglas, E. H. Owens, J. Hardenstine and R. C. Prince, The OSSA II pipeline oil spill: the character and weathering of the spilled oil, *Spill Sci. Technol. Bull.*, 2002, **7**, 135–148.

19. I. R. Kaplan, Age dating of environmental organic residues, *Environ. Forensics*, 2003, **4**, 95–141.

20. S. M. Mudge, Aspects of hydrocarbon fingerprinting using PLS – new data from Prince William Sound, *Environ. Forensics*, 2002, **3**, 323–329.

21. S. M. Mudge, Reassessment of the hydrocarbons in Prince William Sound and the Gulf of Alaska: Identifying the source using partial least squares, *Environ. Sci. Technol.*, 2002, **36**, 2354–2360.
22. G. W. Johnson, R. Ehrlich and W. Full, Principal components analysis and receptor models in environmental forensics, in *Introduction to Environmental Forensics*, ed. B. L. Murphy and R. D. Morrison, Academic Press, London, 2002, pp. 461–515.
23. S. M. Mudge, Multivariate statistics in environmental forensics, *Environ. Forensics*, 2007, **8**, 155–163.
24. K. P. Singh, A. Malik, S. Sinha and V. K. Singh, Multi-block data modeling for characterization of soil contamination: a case study, *Water Air Soil Pollut.*, 2007, **185**, 79–93.
25. J. D. Coop and T. J. Givnish, Gradient analysis of reversed treelines and grasslands of the Valles Caldera, New Mexico, *J. Veg. Sci.*, 2007, **18**, 43–54.
26. S. G. Chu and C. D. Metcalfe, Simultaneous determination of triclocarban and triclosan in municipal biosolids by liquid chromatography–tandem mass spectrometry, *J. Chromatogr. A*, 2007, **1164**, 212–218.
27. B. I. Escher and L. Sigg, Chemical speciation of organics and of metals at biological interphases, in *Physicochemical Kinetics and Transport at Biointerfaces*, ed. H. P. van Leeuwen and W. Köster, Wiley, Chichester, 2004, pp. 205–269.
28. S. M. Mudge and A. S. Ball, Sewage, in *Environmental Forensics: a Contaminant Specific Approach*, ed. R. Morrison and B. Murphy, Elsevier, Amsterdam, 2006, p. 533.
29. G. F. Slater, Stable isotope forensics – when isotopes work, *Environ. Forensics*, 2003, **4**, 13–23.
30. T. C. Schmidt, L. Zwank, M. Elsner, M. Berg, R. U. Meckenstock and S. B. Haderlein, Compound-specific stable isotope analysis of organic contaminants in natural environments: a critical review of the state of the art, prospects and future challenges, *Anal. Bioanal. Chem.*, 2004, **378**, 283–300.
31. J. T. Wilson, R. Kolhatkar, T. Kuder, P. Philp and S. J. Daugherty, Stable isotope analysis of MTBE to evaluate the source of TBA in ground water, *Ground Water Monit. Remed.*, 2005, **25**, 108–116.
32. A. Borja, A. B. Josefson, A. Miles, I. Muxika, F. Olsgard, G. Phillips, J. G. Rodriguez and B. Rygg, An approach to the intercalibration of benthic ecological status assessment in the North Atlantic ecoregion, according to the European Water Framework Directive, *Mar. Pollut. Bull.*, 2007, **55**, 42–52.
33. I. Muxika, A. Borja and J. Bald, Using historical data, expert judgement and multivariate analysis in assessing reference conditions and benthic ecological status, according to the European Water Framework Directive, *Mar. Pollut. Bull.*, 2007, **55**, 16–29.
34. M. E. Arias, J. A. Gonzalez-Perez, F. J. Gonzalez-Vila and A. S. Ball, Soil health – a new challenge for microbiologists and chemists, *Int. Microbiol.*, 2005, **8**, 13–21.

Microbial Techniques for Environmental Forensics

ANDREW S. BALL, JULES N. PRETTY AND RAKHI MAHMUD

1 Introduction

The application of microbiology to environmental forensic investigations involves the application of a range of sub-disciplines including microbial physiology, molecular microbial ecology and microbial biochemistry. Microbial forensics employs a range of techniques to trace a contaminant through the environment using a microbial marker. Many of these techniques, such as selective isolation plating, are well established and have been successfully employed for many years.[1]

The purpose of this chapter is to provide an introduction to the application of traditional microbiological techniques to trace and track environmental contamination. In addition, it provides an introduction to molecular microbial ecology, an emerging sub-discipline within microbiology which has applications in environmental forensics. This review will focus on the benefits of community fingerprinting to the field of environmental forensics, outlining the techniques commonly used and also outlining the potential developments in this emerging field of forensics. Illustrations of the application of these technologies will be presented through examples.

2 Traditional Microbial Forensics

Various traditional microbiological techniques can be used to follow a specific microbial population in the environment. Microbial forensics can be applied to both terrestrial and aquatic environments, although most studies and examples have been based on terrestrial systems. The techniques that can be applied are numerous. However, the basic premise on which these techniques are applied to environmental forensics is that microorganisms are indicators of the contamination event. In a simple example, the presence and fate of faecal contamination in the environment can be followed by determining the number of faecal bacteria (*e.g.* faecal coliforms) in the environmental sample.[2]

Issues in Environmental Science and Technology, No. 26
Environmental Forensics
Edited by RE Hester and RM Harrison
© Royal Society of Chemistry 2008

Microorganisms are generally good indicators of environmental contamination as they are ubiquitous in all environments; however, that is not to say that all bacteria are everywhere. Particular contaminants have an associated microbial community which consists of microorganisms capable of surviving in the presence of the contaminant. It is possible that these organisms have also developed the metabolic capacity to utilize these contaminants, offering the opportunity to remediate the contaminant. Two broad classes of microorganisms associated with the contaminant can be described:

- First, those organisms that were present and constituted part of the contaminant. Faecal contamination is an example of an environmental contaminant which has an associated microflora. In this instance, specific genera of microorganisms are used both to identify and to quantify the level of contamination. In this scenario, it is desirable to monitor a microbial population that is capable of long-term survival in the environment, but which is unable to grow in the environment.
- Second, the monitoring of a microbial population which is present in the environment but which may have not been associated with the contaminant at source, but when released into the environment naturally occurring microorganisms become associated with the contaminant through their utilization of the contaminant. In this instance, the identification of the microorganisms allows identification of the contaminant in the environment but does not quantify any changes in concentration of as contaminant as it moves through an environment. An example would be an oil spill where naturally occurring microorganisms capable of degrading components of the pollutant can be detected.[3,4]

Traditional community profiling techniques useful to environmental forensics include:

- community-level physiological profiling;
- phospholipid fatty acid analysis.

2.1 Community-level Physiological Profiling

Community-level physiological profiling (CLPP) is based on the utilization patterns of individual carbon substrates produced through the use of commercially produced 96-well Biolog microtitre plates. For example, Eco-plates contain 31 different substrates in wells in triplicate. The assay is based on the measurement of the rate of oxidative catabolism of the substrates, indicated through the development of a colour during the substrate utilization. The result of the assay is a pattern of substrate utilization for individual substrates.[5] If the inoculum is an environmental sample then the results are reflective of the potential biological function of the sample. CLPP has been used in a number of studies for examining the potential biological function of an environmental sample. Bossio *et al.*[6] carried out soil microbial community analyses in tropical soils, both agricultural and woodland, using CLPP. The results showed that the greatest substrate diversity was observed in woodland soils. However, the authors concluded that CCLP showed lower specificity with respect to soil type and land usage compared with phospholipid fatty acid analysis or molecular microbial profiling.

2.2 Phospholipid Fatty Acid Profiling

Phospholipid fatty acid analysis (PLFA) is a widely used technique used to detect changes in environmental microbial communities.[7] Changes in PLFA profiles are indicative of changes in the overall structure of microbial communities[8] and signature PLFAs permit further analysis of specific groups of microorganisms present in the community.[8] PLFAs are present in membranes of cells and can be extracted directly from an environmental sample without the need for isolation.[9] Different groups of microorganisms exhibit different PLFA profiles. Gram-negative bacteria contain a higher concentration of mono-unsaturated or cyclic fatty acids.[10] Gram-positive bacteria contain higher concentrations of iso- and anteiso-branched fatty acids.[11] Fungi contain a greater percentage of polyunsaturated fatty acids than bacteria.[12] The sensitivity of the technique was discussed by Bossio *et al.*,[6] who concluded that PLF analysis was capable of revealing shifts in the total soil microbial community in response to different land management regimes.

3 Emerging Microbial Analyses

Molecular environmental forensics can be defined as the application of molecular microbiology to environmental forensics. Molecular environmental forensics provides a means by which a microbial profile is used to trace a contaminant source.[13]

The molecular microbiological fingerprinting techniques that can be applied to the monitoring of these two populations have considerable overlap. This chapter presents an overview of molecular microbial fingerprinting techniques together with examples of their application in environmental forensics.

3.1 Microbial Analysis and Environmental Forensics

Traditional microbiological techniques such as selective isolation plating[14] have been used either to follow a particular microorganism or to follow changes in a microbial community after a contaminant event[15] (Figure 1).

As already discussed, bacteria are generally regarded as good indicators of environmental contamination as they are ubiquitous in nature, being found in all environments, even in extreme conditions such as low pH, high temperature or high salinity.[14] Bacterial communities are also able to assimilate a wide range of contaminating chemicals such as polychlorinated biphenyls (PCBs).[16]

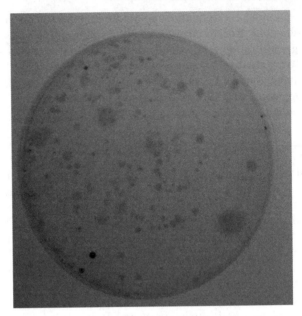

Figure 1   Selective isolation plating of an environmental sample.

Over the past decade, culture-independent techniques have been increasingly used to study microbial communities.[7,17] These techniques generally use DNA and RNA structures to examine the diversity and activity of microbial communities.[18] This approach does not require any microbial isolations and the DNA or RNA extracted from the environmental sample represents the sum of the community DNA.

3.2   The Basis of Molecular Microbial Forensic Techniques

The analysis of DNA represents the most widely used technology in molecular environmental forensics. This section provides an overview of DNA and its application to environmental forensics. DNA is present in almost all known organisms. DNA stores information in genes, discrete sequences of nucleotides. DNA is a polymer consisting of a large repetition of monomeric sequences, called nucleotides. Each nucleotide consists of a deoxyribose (a five-carbon sugar), a nitrogen-containing base and a phosphate group. Deoxyribose and phosphate components are common in all nucleotides, whereas the nitrogen-containing bases may be one of four types. These bases belong to two main classes, purines [adenine (A) and guanine (G)] and pyrimidines [cytosine (C) and thymine (T)] (Figure 2). It is the arrangement of these bases that regulates the production of specific proteins inside the cell. The information in genes is transcribed (written) into ribonucleic acids (RNAs). These contain uracil (U) instead of thymine (T) (AGCU). The 'message' in the RNA is read (translated) and proteins synthesized in ribosomes.

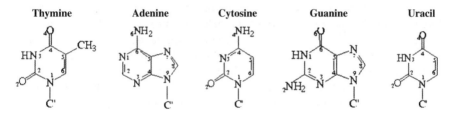

Figure 2 The nucleotides present in DNA and RNA.

DNA represents the basic identity (genotype) of an organism, which in turn determines the physical features (phenotype) of an organism. The DNA profile therefore represents a unique fingerprint that is specific to each individual. In a perfect world, we would manipulate and compare whole genomes to determine relationships among microbes. However, based on current technology, this is not yet feasible due to time and cost constraints. Instead, comparisons between organisms based on phylogenetic markers are used. This is generally a gene whose sequence is used to infer phylogenetic relationships among microbes. A major assumption with this technique is that gene phylogeny more or less reflects the evolutionary history of the microbes possessing the gene of interest.

3.3 Ribosomes

Ribosomes are large, abundant ribonucleoprotein complexes upon which protein synthesis occurs. They are found free in the cytoplasm and in eukaryotic cells, associated with the membranes of the rough endoplasmic reticulum. Ribosomes are the site of protein synthesis. Ribosomes therefore perform a vital function in all cells and are critical to cell function. Ribosomes are present in large numbers in active cells; usually 10 000 to 20 000 ribosomes are present per cell. Ribosomes may therefore occupy up to 25% of the cell volume. Protein synthesis is an energetically demanding process and therefore ribosomes may also utilize up to 90% of the cell's energy.

Ribosomes may be differentiated on the basis of size into large and small subunits. All ribosomes comprise two dissimilar sized subunits, the large and the small subunit that attach to the mRNA at the beginning of protein synthesis and detach when the polypeptide has been translated. Each subunit consists of several ribosomal RNAs (rRNAs) and numerous ribosomal proteins (r-proteins). Their relative size is usually expressed in Svedberg units (S). In *Escherichia coli*, the 70S ribosome is composed of a small 30S subunit and a large 50S subunit. The large subunit comprises 34 proteins and the 23S and 5S rRNAs. The small subunit contains 21 different proteins and the 16S rRNA. In eukaryotic organisms, the ribosomes are larger (80S) with the large subunit (60S) containing 50 proteins and the 28S, 5.8S and 5S rRNAs. The smaller subunit (40S) contains 33 proteins and the 18S rRNA. In Archaea, a prokaryotic form of life that forms, distinct from bacteria and a domain in the tree of life, ribosomes resemble those of bacteria but may contain extra subunits similar to those of eukaryotic cells.

Although bacteria and Archaea appear similar in structure, they have very different metabolic and genetic activity. One defining physiological characteristic of Archaea is their ability to live in extreme environments. They are often called extremophiles and, unlike bacteria and eukarya, depend on either high salt, high or low temperature, high pressure or high or low pH.

As already stated, the function of ribosomes is crucial to the cell; it would therefore be expected that their structural RNAs should not evolve rapidly, as any sequence change may disable the ribosome. Consequently, ribosomal gene sequences are therefore highly conserved, *i.e.* they do not change much over time. It is estimated that the divergence rate for 16S rRNA is 1% per 50 million years, although this estimate may vary by an order of magnitude. The 16S ribosomal RNA therefore represents a universally conserved DNA sequence which is possessed by all bacteria. Also, importantly, with very few exceptions, the 16S ribosomal RNA is not transferred horizontally, that is, that the 16S rRNA is rarely transferred via a process in which an organism transfers genetic material to another cell that is not its offspring. By contrast, vertical transfer occurs when an organism receives genetic material from its ancestor, *e.g.* its parent or a species from which it evolved.

3.4   Ribosomal RNA and Taxonomy

A closer inspection of the 16S ribosomal RNA gene (Figure 3) reveals that the 16S rRNA can fold into a pattern of hairpins and loops that constitute its secondary structure. This folding pattern probably serves as a molecular signpost for allowing recognition of rRNA segments by proteins during assembly of the ribosomal subunits. The 16S rRNA sequence has hypervariable regions, where sequences have diverged over evolutionary time. These are often flanked by strongly conserved regions. The highly conserved secondary structure is useful for detecting polymerase chain reaction (PCR) and sequencing artefacts/errors. However the extent of variation in sequence conservation allows the 16S rRNA to have a broad range of utility in phylogenetic analyses. This was first utilized in the late 1970s, when Woese and colleagues studied the evolutionary relationships among prokaryotes through the comparison of rRNA gene sequences. One of the most important finding of their work was the discovery that not all prokaryotes are related. One group of bacteria, the Archaea, possess rRNA gene sequences that were as unrelated to the Eubacteria as they are to eukaryotes. Some 30 years later, we now routinely investigate phylogenetic relationships between prokaryotes by comparing nucleotide sequences (AGCT) of their 16S (small subunit) rRNA genes.

3.5   Polymerase Chain Reaction (PCR)

In terms of its application to environmental forensics, PCR represents a way of finding a needle in a haystack and subsequently producing a pile of needles from the hay. For example, we may be looking for a specific 300 base pair strand of DNA amongst a sequence of 3 000 000 000 base pairs. The technique,

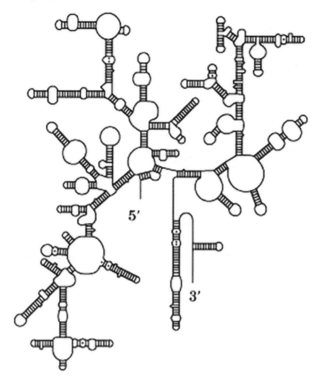

Figure 3 Secondary structure of the 16S ribosomal RNA.

developed in the 1980s, requires only small amounts of sample DNA, in this case DNA isolated from the environment. This makes PCR highly applicable to environmental forensic investigations. To carry out PCR, primers are required. This is a short string of nucleotides (usually 15–30) that are complementary to the first part of the segment of DNA that is being copied. This string of nucleotides, called a primer, attaches to the beginning of the template strand by base pairing. For any target gene, two primers are required to amplify the target; these two primers bind to conserved regions of the rRNA, flanking the target, a variable region of the 16S rRNA that one is trying to amplify. To make primers of the correct sequence that will bind to the template DNA, it is necessary to know a small part of the template sequence on either side of the region of DNA one wishes to amplify. The section of target DNA bounded by the two primers is called the amplicon. The length of the amplicon is variable but ideally should be around 400–600 base pairs.

The PCR process requires three main steps (Figure 4). The first is denaturation of the DNA sample at high temperature (94°C). In this stage, the DNA denatures, splitting the double-stranded DNA into single stands. The second is annealing at around 54°C. In the presence of selected primers, primases and polymerase enzymes are used to identify the target DNA sequence and then

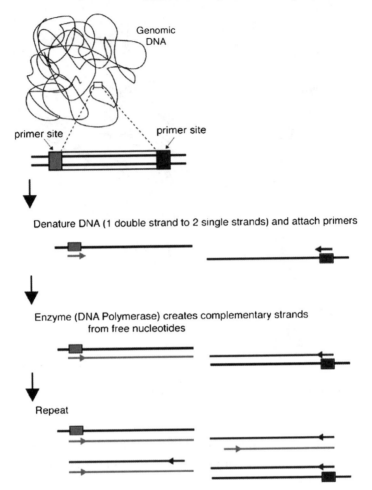

Figure 4   The PCR method.

produce a copy of them. The final step is extension at around 72°C, where bases complementary to the template are coupled to the primer on the 3′ side by polymerases. The cycle is repeated a number of times, with each cycle resulting in a doubling of the number of amplicons. The result of the PCR of DNA extracted from an environmental sample is the production of large numbers of copies of the original template DNA present in the sample. Theoretically, any DNA to which the primers can bind to will be replicated. Therefore, at the end of the PCR reaction, many copies of the same sequence of DNA will be present and subtle differences in the base pair composition of the 16S rRNA sequence will reflect evolutionary divergence between organisms. These differences in base pair composition can be exploited to provide a fingerprint of the community. The methods associated with DNA fingerprinting are discussed below.

4 PCR-based DNA Fingerprinting Techniques

The most common PCR-based community DNA analysis techniques with applicability to environmental forensics include the following:

- denaturing gradient gel electrophoresis (DGGE) and its derivatives;
- single-stranded conformation polymorphisms (SSCP);
- terminal restriction fragment length polymorphisms (TRFLP).

4.1 Denaturing Gradient Gel Electrophoresis and Its Derivatives

Molecular-based fingerprinting based on community DNA and RNA have recently been shown to be an effective technique for the examination of microbial communities.[17] In particular, this technique has been used to study the impact of various environmental factors, including pollutants, climate change and agricultural practice, on the microbial community.[18] The technique is based on the PCR amplification of a short 16S rRNA gene sequence and involves the separation of amplicons of the same length (200–300 base pairs) but with a different base pair sequence on polyacrylamide gels containing a linear gradient of a DNA denaturant. The community fingerprint obtained represents a complex band profile of the genetic structure of the community being investigated.[19] Following gel electrophoresis and staining, bands may be excised from the gel and, following a clean-up procedure, may be sequenced and the identity of members of the community determined.

The factor that allows the separation of the same-sized DNA fragments is the inherent thermal instability of DNA fragments caused by differences in base pair sequences. DGGE allows PCR products of the same length but difference sequence composition to be separated in gradient gels according to the melting behaviour of the DNA. The unzipping of the DNA molecules is achieved when it reaches its critical denaturant concentration, at which point the DNA stops moving. The increasing gradient of denaturant (20–70% in this illustration) which causes the unzipping of double-stranded DNA fragments is obtained through increasing concentrations of formamide and urea. The key steps in DGGE analysis (Figure 5) for community profiling include the following:

- DNA extraction from an environmental sample.
- Amplification of DNA fragments by PCR using primers for the target gene (usually rRNA). Amplicons are designed to include a GC clamp. This is a stretch of GC-rich sequences of 20–36 bp used to introduce a high melting temperature (T_m) domain to each of the target amplicons.
- Incomplete denaturation of the amplicons and separation of the fragments in gels containing a linear gradient of DNA denaturant.
- Staining for visualization of separated fragments.

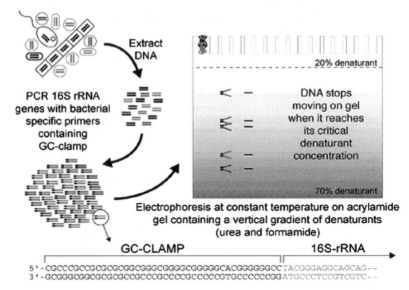

Figure 5 Principal steps in the microbial community profiling by denaturing gradient gel electrophoresis.

Following extraction of DNA from the environmental samples, PCR amplification of the extracted DNA is carried out, usually through the targeting of the 16S rRNA gene. A range of primers targeting the conserved regions of the gene are widely available.[18] The primers used to amplify the 16S rRNA genes contain a GC clamp, which prevents the complete denaturation of DNA fragments. During DGGE, fragments move down the gradient and, with the exception of the GC clamp, denature, halting the mobility of the amplicon. Amplification of the 16S rRNA genes within a community and subsequent analysis by DGGE give rise to a banding pattern in which each band may correspond to a single species.

Temperature gradient gel electrophoresis (TGGE) is a variation of DGGE. Although both are based around the same principle, *i.e.* differentiating DNA based on the thermal properties of different sequences (of the same length), TGGE and DGGE differ in the method used to induce the denaturation of the double-stranded DNA fragment. TGGE uses a linearly increasing temperature gradient in place of an increasing chemical denaturant gradient to achieve the separation of the double-stranded DNA amplicons.

Analyses of PCR-amplified 16S rDNA gene fragments from environmental samples have been widely used in molecular microbial ecology,[17] where they have been employed as a tool to investigate mainly bacterial communities. Figure 6 shows an example of soil DNA profiles illustrating the influence of land management on the temporal changes of a soil microbial community DNA fingerprint. Although similarities in the banding patterns of the soil bacterial communities can be seen, differences in banding patterns over time

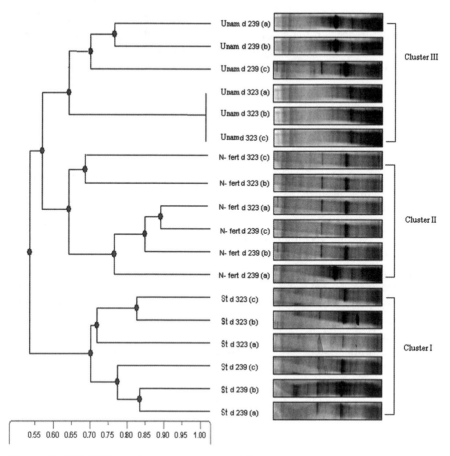

Figure 6 UPGMA dendrogram constructed from similarity matching data (Dice–Sorensen similarity coefficient) produced from DGGE profiles of 16S rDNA amplified from soil samples (collected at d 239 and d 323) amended with straw (represented as St) and N-fertilizer (represented as N-fert) and unamended soil (represented as Unam). The scale bar represents similarity among the triplicate samples.

and with treatment can be observed (Figure 6). Figure 6 also shows the analysis of a data set using unpaired-mean group analysis (UPMGA). Each ribotype (band) was identified and its intensity measured after image capture and analysis using Phoretix ID Advanced software (Non Linear Dynamics, Newcastle, UK). This band intensity was then expressed as a proportion of the total intensity of all of the bands comprising a particular community profile. The software eliminates background and automatically detects peaks when noise levels and minimum peak thresholds are set and was used as described in detail by Girvan *et al.*[17] In this case, the analysis shows that the land management practice, the addition of straw, the addition of N-fertilizer or unamended soil. led to the clustering of the soil microbial community profile. The sampling of soil at different times only showed intra-cluster changes. This analysis was carried out using DGGE, which is an efficient method for the detection of DNA sequence differences and is a convenient tool for analysing changes in a community through analysis of only a small fragment (200–300 base pairs) of the 16S rRNA gene.

4.2   Single-stranded Conformation Polymorphism Analysis (SSCP)

SSCP, like DGGE, distinguishes DNA molecules of the same size, but with different nucleotide sequences. Separation is based on the unique three-dimensional features of single-stranded DNA, which allow small changes in the nucleotide sequence to be detected through conformational changes in the DNA. In the absence of a complementary strand, single-stranded DNA may undergo intrastrand base pairing, resulting in loops that give the DNA its unique three-dimensional structure. In turn, this unique structure imparts a specific motility through a polyacrylamide electrophoresis gel.[20] SSCP analysis requires that DNA is extracted from an environmental sample. PCR is then carried out using one phosphorylated and one non-phosphorylated primer specific for the target gene (usually the 16S rRNA gene). Double-stranded amplicons are then converted to single strands through lambda exonucleases, which digest the phosphorylated strand. The primary steps associated with SSCP are as follows:

- DNA extraction from an environmental sample;
- amplification of DNA fragments by PCR using a phosphorylated and a non-phosphorylated primer for the target gene;
- denaturation of the amplicons to a single-stranded form by exonuclease digestion;
- separation of the denatured amplified fragments using polyacrylamide gel electrophoresis;
- visualization of the separated fragments by silver staining or autoradiography.

Using SSCP, community patterns can be obtained from a range of environmental samples. Following visualization, bands can be excised, re-amplified by PCR and sequenced directly. SSCP therefore represents a low-cost methodology for the analysis of the diversity of a microbial community in an environmental sample. However, single-stranded DNA mobility is dependent not only on the unique three-dimensional structure of the amplicon, but also on temperature and pH. Therefore, it is better to run gels at constant temperature and low pH. Also, for optimal results, the DNA fragment size should fall within the range 150–300 base pairs.

SSCP represents a simple, low-cost technique for the analysis of the diversity of a microbial community in environmental samples, based on PCR-amplified small subunit (SSU) rRNA gene sequences from DNA extracted from environmental samples.

4.3 Terminal Restriction Fragment Length Polymorphism (TRFLP)

TRFLP represents one of the most common DNA fingerprinting techniques in environmental forensic applications. TRFLP is a microbial community profiling method usually based around the 16S rRNA gene. TRFLP is again based on PCR amplification of the target gene, but uses a fluorescent end-labelled primer. Following amplification, amplicons are digested using restriction endonucleases with high specificity. Fragments (different sizes) are separated by electrophoresis with the visualization of only the terminal fragments as they contain the fluorescent label. These amplified fragments of DNA originate from different organisms and consequently have sequence variations, ensuring that terminal restriction sites for different species in a community are unique. TRFLP uses the differences in length from different DNA terminal fragments to differentiate between profiles of microbial communities.

The main steps of TRFLP analysis are as follows (Figure 7):

- DNA extraction from an environmental sample;
- amplification of DNA fragments by PCR using fluorescently labelled primers;
- digestion of amplicons with one or more restriction endonucleases;
- separation and visualization of fluorescently labelled terminal fragments.

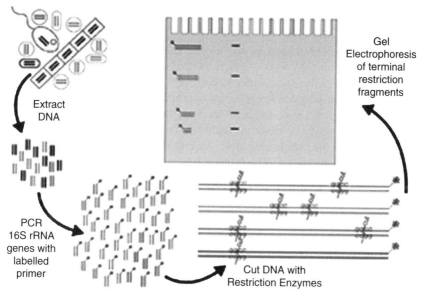

Figure 7 Principal steps in the microbial community profiling by terminal restriction fragment length polymorphism.

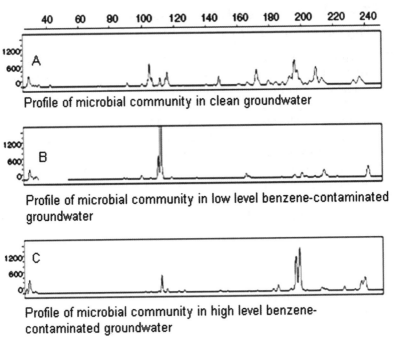

Figure 8 Electropherograms showing bacterial 16S rDNA TRFLP profiles of *in situ* communities from three benzene-contaminated groundwater wells containing either clean groundwater (A), low-level benzene contamination (B) or high-level benzene contamination (C). The horizontal scale represents the TRF length in nucleotides and the vertical scale the relative fluorescence (redrawn from Fahy *et al.*[3]).

TRFLP patterns are routinely used to characterize the microbial communities from sites contaminated with a pollutant and comparison with the microbial community profile from uncontaminated areas of the site. Figure 8 illustrates how TRFLP profiles distinguish between the microbial communities of groundwater samples contaminated with different levels of benzene.[3] The profile of the microbial community in the clean groundwater sample (A) contains a number of peaks throughout the profile indicative of a complex community. However, in the presence of benzene (B and C), profiles become simpler with fewer peaks, indicative of a community dominated by fewer organisms, presumably as a result of the benzene contamination. The relative intensity of any single peak provides some information regarding the concentration of the particular microorganism present in the community, with large peaks indicating a greater prevalence of that organism. TRFLP is a technique that is capable of rapidly analysing large amount of information through automation, resulting in the production of large quantities of reproducible data. TRFLP enables communities to be monitored at high resolution, especially when TRFLP is linked to a capillary electrophoresis sequencer, permitting increased throughput and greater reproducibility.[21] One specific disadvantage of TRFLP lies in the complexity associated with identification of organisms responsible for a particular element in a profile. This is because TRFLP involves destructive sampling; the DNA cannot be recycled. This is in contrast to DGGE technology, which allows for either direct cloning of bands or indeed the direct sequencing of bands excised from the gel.

5   Limitations of PCR-based Methodologies

PCR technologies have limitations and, although some of these limitations are specific to the technique used (Table 1), there are some general caveats that can be made regarding the use of PCR:

- PCR primers are only able to amplify genes that match the primer sequence.
- PCR amplification can be uneven; some sequences are readily amplified but others less so.
- Under-amplification of sequences from abundant taxa is possible, skewing the community fingerprint.
- The results of studies based on these techniques may be regarded as an inventory of the community rather than a quantitative measurement of abundant taxa.
- The outcome illustrates the PCR-amplifiable community in the sample.

Another perceived limitation in terms of environmental forensics with PCR based on DNA is that this technology does not differentiate between DNA from living and dead organisms. DNA from any organisms present in the sample may be amplified. This may be important when investigating organisms capable of living in a contaminated sample, for example when considering the application of bioremediation. In this instance, it is possible to extract RNA not DNA from an environmental sample. RNA is only present in living organisms. In terms of tracking pollution, non-discrimination between living and dead microorganisms may be useful if the microbial community of the original material is to be traced through varying communities (e.g. faecal pollution). In contrast, examination of a contaminant plume through an environment may require the identification of living microorganisms which are used to track the contaminant plume.

6   Forensic Interpretation of Profiles

Ideally the various DNA techniques that are used to generate community profiles should be analysed similarly. As we have seen, similarity indices such as the Dice–Sorenson similarity coefficient can be used.[23,24] However Horswell *et al.*[23] suggested that such an index may not be robust enough to be used as evidence in a court of law. Clearly, high similarity indices between replicates and low similarity

Table 1   Summary of the commonly used 16sRNA community fingerprinting techniques used in environmental forensics.

Technique	Mode of differentiation	Advantages	Disadvantages	Reference
Denaturing gradient gel electrophoresis (DGGE)	Melting temperature of DNA in a denaturing gradient	Bands can be excised from gel and sequenced	Different sequences can have similar melting properties	19
		Probes can be used to hybridize to profile	Only small PCR products (2–300 bp) can be separated efficiently	
Temperature gradient gel electrophoresis (TGGE)	Melting temperature of DNA in a temperature gradient	Bands can be excised from gel and sequenced	Different sequences can have similar melting properties	22
		Probes can be used to hybridize to profile	Only small PCR products (2–300 bp) can be separated efficiently	
Single-stranded conformation polymorphism (SSCP)	Secondary structure of single-stranded DNA	Bands can be excised from gel and sequenced	Electrophoretic conditions are important variables. Only small PCR products (150–400 bp) can be used	
Terminal restriction length fragment polymorphism (TRFLP)	Terminal restriction fragment length	Automation is well established. Reproducible technology allowing comparisons with other data	Fragments cannot be sequenced directly	21

indices for different samples are to be preferred. One further factor to consider is the heterogeneity of soils even over very short distances.[25] However, less is known about the variability in the microbial community over such distances. Felske and Akkermans[26] showed through the use of TGGE that soil samples collected 1 m apart contained the same prominent bacteria.

7 Conclusions

The requirement for analysis of both aquatic and terrestrial samples in cases involving environmental forensics is becoming increasingly appreciated. However, because of the limitations of the currently available techniques, this analysis is rarely used as evidence. The recent developments in the application of molecular biology have provided for the first time effective tools to examine and compare the microbial community (through DNA fingerprinting) of environmental samples. These DNA profiling techniques are based around the extraction of DNA (and RNA) directly from the environmental sample. The three techniques described here (DGGE/TGGE, SSCP and TRFLP) profile PCR-amplified genes from microbial community (mainly bacterial) DNA, through targeting of the 16S rRNA gene, resulting in the generation of a fingerprint of the microbial community. Each of the methods described have specific advantages and disadvantages, although all three are subject to the limitations associated with PCR. However, DNA fingerprinting of environmental samples offers great potential as a tool in environmental forensics.

References

1. A. S. Ball, Bacterial cell culture, in *Encyclopedia of Molecular Cell Biology and Molecular Medicine*, Wiley, Chichester, 2004.
2. S. M. Mudge and A. S. Ball, *Sewage*, in *Environmental Forensics – Contaminant Specific Guide*, eds. R. D. Morrison and B. L. Murphy, Academic Press, New York, 2006, pp. 36–53.
3. A. Fahy, G. Lethbridge, R. Earle, A. S. Ball, K. N. Timmis and T. J. McGenity, Effects of long-term benzene pollution on bacterial diversity and community structure in groundwater, *Environ. Microbiol.*, 2005, **7**, 1192–1199.
4. A. Fahy, T. J. McGenity, K. N. Timmis and A. S. Ball, Heterogeneous aerobic benzene-degrading communities in oxygen-depleted groundwaters, *FEMS Microbiol. Ecol.*, 2006, **58**, 260–270.
5. J. L. Garland and A. L. Mills, Classification and characterization of heterotrophic microbial communities on the basis of patterns of community-level-sole-carbon-source-utilization, *Appl. Environ. Microbiol.*, 1991, **57**, 2351–2359.
6. D. A. Bossio, M. S. Girvan, J. Bullimore, A. S. Ball, V. Verchot, T. B. Moses, K. M. Scow, J. N. Pretty and A. M. Osborn, Soil microbial community response to land use change in an agricultural landscape in western Kenya, *Microbial Ecol.*, 2005, **49**, 50–62.
7. R. Turpeinen, T. Kairesalo and M. M. Häggblom, Microbial community structure and activity in arsenic-, chromium- and copper-contaminated soils, *FEMS Microbiol. Ecol.*, 2004, **47**, 39–50.
8. A. Frostegard, A. Tunlid and E. Baath, Changes in microbial community structure during long-term incubation in two soils experimentally contaminated with minerals, *Soil Biol. Biochem.*, 1996, **28**, 55–63.

9. T. Pennanen, A. Frostgard, H. Fritze and E. Baath, Phospholipid fatty acid composition and heavy metal tolerance of soil microbial communities along two heavy metal polluted gradients in coniferous forests, *Appl. Environ. Microbiol.*, 1996, **62**, 420–428.

10. S. G. Wilkinson, Gram-negative bacteria, in *Microbial Lipids*, Vol. 1, eds. C. Ratledge and S. G. Wilkinson, Academic Press, London, 1988, pp. 299–488.

11. W. M. O'Leary and S. G. Wilkinson, Gram-positive bacteria, in *Microbial Lipids*, Vol. 1, eds. C. Ratledge and S. G. Wilkinson, Academic Press, London, 1988, pp. 117–202.

12. T. W. Federle Microbial distribution in soil – new techniques, in *Perspectives in Microbial Ecology*, F. Megusar and M. Gantar, eds., Slovene Society for Microbiology, Ljubljana, 1986, pp. 493–498.

13. I. G. Petrisor, R. A. Parkinson, J. Horswell,J. M. Waters, L. A. Burgoyne, D. E. A. Catcheside, W. Dejonghe, N. Leys, K. Vanbroekhoven, P. Pattnaik and D. Graves, Microbial forensics, in *Environmental Forensics – Contamination Specific Guide*, eds. R. D. Morrison and B. L. Murphy, Academic Press, New York, 2006, pp. 227–257.

14. A. S. Ball, *Bacterial Cell Culture – Essential Data*, Wiley, Chichester, 1997, p. 100.

15. B. Budowle, S. E. Schutzer, A. Einseln, L. C. Kelley, A. C. Walsh, J. A. Smith, B. L. Marrone, J. Robertson and J. Campos, Building microbial forensics as a response to bioterrorism, *Science*, 2003, **301**, 1852–1853.

16. H. G. Truper, Prokaryotes – an overview with respect to biodiversity and environmental importance, *Biodiversity Conserv.*, 1992, **1**, 227–236.

17. M. S. Girvan, J. Bullimore, J. N. Pretty, A. M. Osborn and A. S. Ball, Soil type is the primary determinant of the composition of the total and active bacterial communities in arable soils, *Appl. Environ. Microbiol.*, 2003, **69**, 1800–1809.

18. M. S. Girvan, J. Bullimore, A. S. Ball, J. N. Pretty and A. M. Osborn, Monitoring of seasonal trends in the soil microbial community of an agricultural field, *Appl. Environ. Microbiol.*, 2004, **70**, 2692–2701.

19. G. Muyzer, E. C. Dewaal and A. G. Uitterlinden, Profiling of complex microbial populations by denaturing gradient gel electrophoresis analysis of polymerase chain reaction amplified genes coding for 16S ribosomal RNA, *Appl. Environ. Microbiol.*, 1993, **59**, 695–700.

20. U. Melcher, SSCPs. http://opbs.okstate.edu/~melcher/MG/MGW1/MG11129.html. Accessed 17August 2007.

21. A. M. Osborn, E. R. Moore and K. N. Timmis, An evaluation of terminal restriction fragment length polymorphism (T-RFLP) analysis for the study of microbial community structure and dynamics, *Environ. Microbiol.*, 2000, **2**, 39–50.

22. H. Heuer, M. Krsek, P. Baker, K. Smalla and E. M. Wellington, Analysis of actinomycete communities by specific amplification of genes encoding 16S rRNA and gel-electrophoretic separation in denaturing gradients, *Appl. Environ. Microbiol.*, 1997, **63**, 3233–3241.

23. J. Horswell, S. J. Cordiner, E. W. Mass, B. W. Sutherland, T. W. Speier, B. Nogales and A. M. Osborn, Forensic comparison of soils by bacterial community DNA profiling, *J. Forensic Sci.*, 2002, **47**, 350–353.

24. C. B. Blackwood, T. Marsh, S. H. Kim and E. U. Paul, Terminal restriction fragment length polymorphism data analysis for quantitative comparison of microbial communities, *Appl. Environ. Microbiol.*, 2003, **69**, 926–932.

25. J. I. Prosser, *Microbial processes within the soil*, in *Modern Soil Microbiology*, eds. J. D. van Elsas, J. T. Trevors and E. M. Wellington, Marcel Dekker, New York, 1997, pp. 183–203.

26. A. Felske and A. D. Akkermans, Spatial homogeneity of abundant bacterial 16S rRNA molecules in grassland soils, *Microbial Ecol.*, 1998, **36**, 31–36.

Spatial Considerations of Stable Isotope Analyses in Environmental Forensics

JAMES R. EHLERINGER, THURE E. CERLING, JASON B. WEST, DAVID W. PODLESAK, LESLEY A. CHESSON AND GABRIEL J. BOWEN

Stable isotope analyses complement other analytical approaches to chemical identification in an environmental investigation[1] (*e.g.* HPLC, GC–MS, LC–MS), because stable isotope analyses provide an additional 'fingerprint' that further characterizes a piece of forensic evidence. The analysis of stable isotope composition of a material at natural abundance levels provides a means of relating or distinguishing two pieces of evidence that have exactly the same chemical composition.[2,3] The different kinds of materials, approaches and applications of stable isotope analyses to the forensic sciences have been recently reviewed.[2,4–6] Here we focus on a relatively new and different aspect of applying stable isotopes to forensic sciences: sourcing of evidentiary materials with a geo-spatial perspective.

The study of stable isotopes as an environmental forensic tool is based on the ability of an isotope ratio mass spectrometer to measure very small, naturally occurring differences in the abundances of the heavy (rarer) to light (common) stable isotopes in a material and then to relate that isotope ratio composition to other samples or other evidence (elaborated further below). The specific instrumentation with applications considered in this review is the gas isotope ratio mass spectrometer (IRMS), which analyzes gases with a mass of <70.[7] This limits the applications of the IRMS to analyses of the stable isotope composition of hydrogen (H) as H_2, carbon (C) as CO_2, nitrogen (N) as N_2, oxygen (O) as CO, CO_2 or O_2, chlorine (Cl) as CH_3Cl and/or sulfur (S) as SO_2. Evidence or materials may originate in gaseous, liquid or solid states and the materials analyzed vary from simple to complex with high molecular weights. Whenever the evidence or material to be analyzed is complex, a combustion or pyrolysis reaction is coupled prior to the mass spectrometer to convert the materials into

Issues in Environmental Science and Technology, No. 26
Environmental Forensics
Edited by RE Hester and RM Harrison
© Royal Society of Chemistry 2008

one of the mentioned gases. In some applications, compound-specific analyses are conducted (*e.g.* *n*-alkanes contained within oil), whereas in others complex biological tissues (*e.g.* bird feathers, hair) and whole organisms are analyzed. For isotope ratio analyses of heavier elements and elements difficult to maintain in a gaseous state, a thermal ionization mass spectrometer (TIMS) is employed. The reader is encouraged to look at recent reviews where the application of TIMS approaches to forensic and sourcing applications has been considered.[8,9]

In forensic studies, investigators often compare individual samples with each other to evaluate the extent to which the samples have overlapping to identical chemical compositions. Stable isotopes add another dimension to these sample comparisons by providing an additional isotopic fingerprint that allows materials of similar chemical composition to be distinguished or to be related.[2,5] When sufficient specimens of materials of a given type and of known origin have been acquired and examined, this database may be useful in assigning a probable geographic region-of-origin to a material based simply on an accumulation of observations, rather than to a first-principles mechanistic basis. Such databases have been accumulated for controlled substances such as marijuana, heroin and cocaine.[10–14] These approaches have been useful in both providing information regarding possible region-of-origin sources of controlled substances and in eliminating other regions as possible source locations.

A recent example that highlights the need for spatial maps is the analysis of heroin specimens seized from a North Korean freighter.[12] Australian police seized 50 kg of heroin hydrochloride from the freighter and another 75 kg of heroin hydrochloride from the offload site on Australian soil. Authorities believed that it was 'highly likely' that North Korea was dealing in illegal drugs. In this case, seized heroin specimens were analyzed and determined to be of unknown origin and could not be assigned a probable source location by comparison with other observations of known authentics in the Drug Enforcement Administration (DEA) database. While such an approach can be used to identify samples of unknown origin, assuming that the isotopic composition of a biologically based material from a geographic region maintained a similar isotopic composition over time (different production periods), as has been with natural plant systems,[15–17] the approach provides no forward-looking perspective on the sample's origins.

A novel approach to geographical assignment may yield greater information and insight, especially in the case when an incomplete database is available. By using first-principles or other models of stable isotope fractionation in organisms and the spatial modeling capacity of geographic information systems (GIS), spatial maps of predicted stable isotope ratios can be constructed.[4,18,19] Analytical results from a specimen can then be compared with the map predictions and potential source locations can be identified. These advances represent cutting edge applications of stable isotope ratios to forensics that will continue to develop over the coming years. As databases and fundamental understanding of stable isotope ratios grow, so will their capacity to aid in forensic work. Here we review the potential of this approach to some aspects of the sourcing of materials of forensic interest.

1   A Background in Stable Isotopes

1.1   Stable Isotopes – a Primer

Different isotopes of an element are based on the numbers of neutrons within the nucleus. Stable isotopes are those isotopes of an element that do not decay through radioactive processes over time. Most elements consist of more than one stable isotope. For instance, the element carbon (C) exists as two stable isotopes, ^{12}C and ^{13}C, and the element hydrogen (H) exists as two stable isotopes, ^{1}H and ^{2}H (also known as deuterium). Table 1 provides the average stable isotope abundances of the elements analyzed by IRMS investigations.

1.2   Isotope Ratio Composition is Presented in Delta Notation

Stable isotope contents are expressed in 'delta' notation as δ values in parts per thousand (‰), where δ‰ $= (R_s/R_{Std} - 1) \times 1000$‰ and R_s and R_{Std} are the ratios of the heavy to light isotope (*e.g.* ^{13}C/^{12}C) in the sample and the standard, respectively. We denote the stable isotope ratios of hydrogen, carbon, nitrogen, oxygen and sulfur in delta notation as δ^2H, δ^{13}C, δ^{15}N, δ^{18}O and δ^{34}S, respectively.

R values have been carefully measured for internationally recognized standards. The standard used for both H and O is Standard Mean Ocean Water (SMOW), where $(^2$H/^{1}H$)_{std}$ is 0.0001558 and ^{18}O/^{16}O is 0.0020052. The original SMOW standard is no longer available and has been replaced by a new International Atomic Energy Agency (IAEA) standard, V-SMOW. The international carbon standard is PDB, where $(^{13}$C/^{12}C$)_{std}$ is 0.0112372 and is based on a belemnite from the Pee Dee Formation. As with SMOW, the original PDB standard is no longer available, but the IAEA provides V-PDB with a similar R value. Atmospheric nitrogen is the internationally recognized standard with an R value of $(^{15}$N/^{14}N$)_{std}$ of 0.0036765. Lastly, the internationally recognized standard for

Table 1   Abundances of stable isotopes of light elements typically measured with an isotope ratio mass spectrometer.

Element	Isotope	Abundance (%)
Hydrogen	^{1}H	99.985
	^{2}H	0.015
Carbon	^{12}C	98.89
	^{13}C	1.11
Nitrogen	^{14}N	99.63
	^{15}N	0.37
Oxygen	^{16}O	99.759
	^{17}O	0.037
	^{18}O	0.204
Sulfur	^{32}S	95.00
	^{33}S	0.76
	^{34}S	4.22
	^{36}S	0.014

sulfur is CDT, the Canyon Diablo Troilite, with a value of $(^{34}S/^{32}S)_{std}$ of 0.0450045. Typically, during most stable isotope analyses, investigators would not use IAEA standards on a routine basis. Instead, laboratories establish secondary reference materials to use each day that are traceable to IAEA standards and that bracket the range of isotope ratio values anticipated for the samples.

1.3 Gas Isotope Ratio Mass Spectrometer

The instrumentation with applications considered in this review is the gas isotope ratio mass spectrometer, which analyzes gases with a mass of <70.[1] High-precision measurements of the stable isotope abundance in a known compound or material are made by converting that substance into a gas and introducing the gas into the mass spectrometer for analysis. Combustion or pyrolysis of the sample is usually associated with an elemental analyzer and the gases are separated from each other through a GC interface. At the inlet to the mass spectrometer, the purified gas is ionized and the ion beam is then focused and accelerated down a flight tube where the path of the ion species is deflected by a magnet. Based on the different isotopic compositions of the ions, they are differentially deflected by the magnet. For instance, for measurements of the carbon isotope composition of a material, the carbon is oxidized to produce CO_2 and the primary species formed are $^{12}C^{16}O^{16}O$ (mass 44), $^{13}C^{16}O^{16}O$ (mass 45) and $^{12}C^{18}O^{16}O$ (mass 46). In contrast to a traditional mass spectrometer where the strength of the magnet is varied and the ionic species are measured by a single detector, in the isotope ratio mass spectrometer the magnet field is fixed and the different isotopic ionic species are deflected into separate detector cups at the end of the flight tube, allowing for greater sensitivity in the measurement of the $^{13}C/^{12}C$ ratio. Whereas a traditional mass spectrometer may be able to detect a 0.5% difference in the R_s value of $^{13}C/^{12}C$ in a sample such as might occur in ^{13}C-enriched biochemical studies, an isotope ratio mass spectrometer has the capacity to resolve a 0.0002% difference in the R_s value at the low end of the naturally occurring $^{13}C/^{12}C$ range. In the case of CO_2, this is because R_s is measured in an isotope ratio mass spectrometer as the ratio of the simultaneous currents in the two cups: $^{12}C^{16}O^{16}O$ (mass 44) and $^{13}C^{16}O^{16}O$ (mass 45) detectors.

An elemental analyzer, or gas chromatograph, is often coupled to the front end of the isotope ratio mass spectrometer. With such an arrangement, it is then possible to oxidize the sample for C, N or S isotope analyses using the elemental analyzer, separate the different combustion gases using a gas chromatograph and analyze the sample gases as they pass into the inlet of the isotope ratio mass spectrometer (so-called continuous-flow IRMS). In this case, helium is used as a carrier gas, transporting the combustion gas products from the elemental analyzer to the mass spectrometer. Similarly, it is possible to pyrolyze the sample in the elemental analyzer for hydrogen and oxygen isotope ratio analyses. In addition, samples can also be isolated, purified and combusted offline and then introduced into the mass spectrometer for stable isotope analyses. Reference materials of known isotopic composition are analyzed before or after the sample is analyzed. This improves the accuracy of a measurement by directly comparing the isotope signals of the sample and the working reference materials with each other. As a consequence, a daily calibration is not used with the instrument because essentially every sample is compared with a standard treated to the same preparation and analysis conditions. These working reference materials are identified by the stable isotope community and are exchanged among different laboratories in order to determine the best estimate of the actual stable isotope composition of the working standard. Under the best conditions, there will be many working reference materials reflecting a range of stable isotope ratios and a range of material types (*e.g.* water, plant material, animal tissue).

2   The Stable Isotopes of Water

Most of the water actively involved in today's water cycle is in the oceans, which have globally averaged δ^2H and $\delta^{18}O$ values near 0‰ each on the International SMOW scale. As surface ocean waters evaporate into the atmosphere, the clouds formed are isotopically depleted in ^2H and ^{18}O relative to the ocean, resulting in an air mass that is isotopically depleted relative to the ocean.[20] In turn, as moisture is condensed from clouds during precipitation events, that water is isotopically enriched in ^2H and ^{18}O relative to the cloud, leaving the residual cloud mass isotopically depleted in ^2H and ^{18}O relative to the original cloud mass. The process of differential isotope depletion during precipitation results in a predictable pattern of depleted isotope ratios of precipitation as cloud masses move inland.[20] Since the hydrogen and oxygen in precipitation become the primary source of H and O atoms incorporated into carbohydrates, proteins and lipids during microbe, plant and animal growth, these H and O isotopes have the potential to carry a geographically based piece of information that proves to be useful in forensic studies.

2.1   The Meteoric Water Line

The predictable relationship between the δ^2H and $\delta^{18}O$ values of precipitation is known as the meteoric water line (MWL) and has a globally averaged $\delta^2H/\delta^{18}O$ slope of approximately 8 and an intercept of $+10‰$.[20] Precipitation falling near coastal regions on every continent has δ^2H and $\delta^{18}O$ values slightly less than 0‰. Successive precipitation events associated with rain out of the residual cloud mass result in precipitation with δ^2H and $\delta^{18}O$ values near -200 and $-24‰$ on the continental interiors, respectively. The $\delta^2H/\delta^{18}O$ slope of precipitation is temperature sensitive, but for temperate and tropical regions of the world the slope remains near 8.

2.2   Isotopes of Water on a Spatial Scale

Over half a century ago, the International Atomic Energy Agency (IAEA) in Vienna established a global network of precipitation sites where the δ^2H and $\delta^{18}O$ values of precipitation were monitored on a monthly basis.[21,22] From these and other observations, both temporal and spatial patterns emerged that established the basis for a global representation of the distribution of isotopes in water on a global basis.

Bowen and colleagues have been able to extrapolate from the available location-specific data of stable isotopes in water to spatial maps of the predicted isotopic composition of water throughout the world.[4,23-25] Analyses of the spatial distributions of hydrogen isotopes of waters across North America and Europe reveal substantial variations in stable isotope ratios (Figure 1), making it possible to distinguish precipitation in many geographical locations on the basis of their δ^2H and $\delta^{18}O$ values. There are no unique stable isotope ratio values for waters in a specific geographic location, but rather gradients or bands of different isotope ratio values allowing one to distinguish between locations if they were sufficiently far apart from each other. Today it is possible to estimate reliably the globally averaged δ^2H and $\delta^{18}O$ values of precipitation for different latitudes and longitudes using a calculator available at http:// waterisotopes.org. The spatial integration of these calculations is the map shown in Figure 1. This map of incoming precipitation values is appropriate for use in addressing the stable isotope variations of plants, animals and microbes across the landscape that use this water source during their growth. A second and slightly different water map is generated when tap waters are measured.[19,26] Such a map differs because of water transportation and storage issues, each of

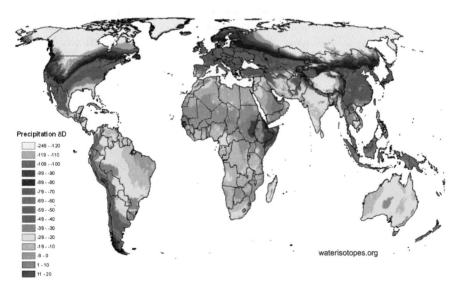

Precipitation δD

	-249 - -120
	-119 - -110
	-109 - -100
	-99 - -90
	-89 - -80
	-79 - -70
	-69 - -60
	-59 - -50
	-49 - -40
	-39 - -30
	-29 - -20
	-19 - -10
	-9 - 0
	1 - 10
	11 - 20

waterisotopes.org

Figure 1 Predicted long-term annual average precipitation hydrogen isotope ratios for the land surface. This continuous layer is produced with a combination of empirical relationships between measured precipitation δ^2H and latitude and elevation and a geostatistical smoothing algorithm for variations not explained by that relationship. Measured precipitation values are those maintained in the IAEA water isotope database. Methods and grids are available at http://waterisotopes.org.

which can result in δ^2H and $\delta^{18}O$ values that are similar to the values expected from incoming precipitation or somewhat more enriched or depleted.

3   Spatial Forensic Applications Based on H and O Isotopes

Natural isotope fractionation processes and fractionation steps associated with the synthesis of biological products lead to a wide range of stable isotope ratio values in plants, animals and microbes.[27–31,3] Since these biological materials are based on the isotopic composition of water as a substrate, to a first approximation we anticipate a strong correlation between the isotope ratios of precipitation in a region and those of organisms living in that region. However, the actual δ^2H and $\delta^{18}O$ values of the carbohydrates, proteins and lipids in biological organisms will differ from those of their water source because of fractionation events associated with the physical environment and fractionation processes during biosynthesis.

In this section, we consider three classes of biologically derived materials that might be analyzed in a forensic case: plant products used in manufacturing, food products and animal systems. For each, we consider relationships between the isotope ratio of precipitation as a measure of geographic location and the fractionation events, which become permanently recorded in the organic matter of biological tissues. Each approach is built on the relationships among the stable isotope ratio of water at a location, statistically derived climatic conditions at that location and the mechanistic relationships relating fractionation events between substrate and product. Since there is often a wide range of δ^2H and $\delta^{18}O$ values in biological materials, this provides an opportunity both to predict and to detect stable isotope ratio differences among samples of forensic, anthropological, ecological and economic interest.

3.1   Cotton as an Example of Plant Sourcing

Plant fibers are of fundamental importance to all of us today, appearing in countless products used on a daily basis. Some of these widespread fiber applications include paper, packaging, construction and clothing. Consider cotton, one of the most important fiber-producing plants and widely grown around the world (Figure 2). Identifying the origins of cotton fibers poses a challenge to the forensic scientist interested in, for example, tracing the origins of a document printed on paper containing cotton fibers or to commercial parties interested in determining if Egyptian cotton bedding had been produced in Egypt or cultivated elsewhere in the world. Here, isotope ratio analyses may offer insights, because spatial patterns in the isotope ratios of cotton are predictable and distinguishable on global scales.

Cotton fibers are composed of cellulose and the hydrogen and oxygen isotope ratios of cellulose molecules are predictable based on mechanistic models.[31] Cellulose contains hydrogen and oxygen atoms derived from water during plant growth and subsequently influenced by atmospheric conditions and biochemical fractionation processes.[31–35] The same mathematical approach allows predictions

Predicted cotton boll δ¹⁸O

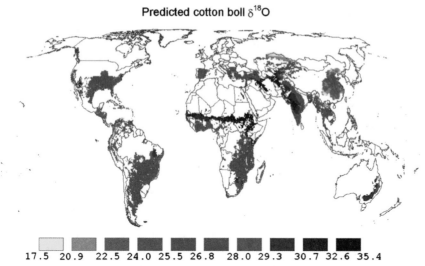

17.5 20.9 22.5 24.0 25.5 26.8 28.0 29.3 30.7 32.6 35.4

Figure 2 A GIS-based map of the predicted oxygen isotope ratio variations of cellulose in cotton fibers expected on the basis of geographical variations in the isotope ratio of water and of the climatic conditions. Unpublished data from J. Ehleringer and J. West.

of the hydrogen and oxygen isotope ratios of cellulose produced in different plant parts: cotton bolls (a reproductive structure), jute (derived from leaf cellulose) or wood (after extracting cellulose from non-cellulose components).

The models used to predict cellulose isotope ratios in plants are utilized to generate spatial predictions using geographic information system approaches. West *et al.*[36] incorporated the mathematics of cellulose isotope fractionation models for both the organic hydrogen and oxygen atoms contained within cellulose. Combining this mathematical approach with grid-cell estimates of both the stable isotope composition of water and the appropriate climate statistics, it is possible to make predictions of the anticipated hydrogen and oxygen isotope ratios for cellulose produced in different parts of the globe. By overlaying the growth locations of cotton with predicted cellulose isotope values, it is possible to derive a global map of the predicted hydrogen and oxygen isotope ratios of cotton fibers (Figure 2).

One advantage of a global map of the hydrogen or oxygen isotope ratios of cotton fibers is that it allows an application common to forensic science: is the evidentiary material consistent with or not consistent with the value of a sample from a specific region? This allows an investigator either to eliminate a location from further consideration or to use the positive result in a corroborative manner when combined with other supporting information. In few cases will there ever be uniquely predicted hydrogen and oxygen isotope ratios for materials such as cellulose. This is simply because of environmental overlap in homologous climatic regions of the world.

One application is analyses of the geographical origins of high-quality counterfeits of security documents, such as currency. Since the mid-1990s, there has been an ongoing investigation by the US Secret Service to determine the origins of the 'Supernote', a very high-quality counterfeit $100 bill.[37,38] Reports have suggested that these 'Supernotes' originated from the Middle East through North Korea. One can imagine that isotope ratio analyses of the fibers in these counterfeit currencies might allow three points to be clarified. Are the counterfeit currency specimens consistent with (a) currency paper currently in use by the US government, (b) cotton fibers that are produced in different parts of the Middle East and (c) cotton fibers that are produced in different parts of North Korea? The answers to these questions are of forensic interest in tracing the origins of these high-quality counterfeit notes.

3.2   Wine as an Example of Food Sourcing

Hydrogen and oxygen stable isotope ratios have been explored as useful markers of geographic source, production methods and even vintages of wines.[39–42] The methods currently in use require large databases of authentic samples for comparison with suspect samples. Identifying or verifying the geographic source of wine can also be addressed with a complementary approach that is based on model descriptions of the spatial variation in expected wine isotope ratios over a large region. These model descriptions can be developed based on first-principles (*e.g.* the biophysics of isotope fractionation in wine grapes) or on regression or other model-fitting approaches.

The regression approach has been used to model successfully wine oxygen isotope ratio variation using GIS.[43] We extend the spatial extent of this model here for all of Washington, Oregon and California (Figure 3) to demonstrate the potential power of the regression/GIS approach. This GIS-based extrapolation allows one to predict the expected values for regions not yet sampled or values anticipated for a region, were grapes to be grown in that region at some time in the future. The approach also provides a basis for predicting the anticipated isotope ratios of mixtures, where grape juices were derived from different geographical regions.

Some combination of database development and spatial modeling of wine stable isotope ratios will continue to be an important tool as consumer demand for terroir (loosely a sense of place) increases. The use of the database approach has been successful in the European Union and will likely continue to be an important source of verification using stable isotope ratios of wine. Emerging approaches that take advantage of spatial tools such as GIS and improved mechanistic understanding of fractionating processes in grapes and the vinification process will continue to enhance the utility of stable isotopes in geographic sourcing of wine.

3.3   Keratin as an Example of Animal Sourcing

Geographic variations in the hydrogen and oxygen isotope ratios of water also form a basis of a geographic signal recorded in the hydrogen and oxygen isotope ratios of organic matter in animals, such as humans,[29,44–46] birds,[47,48] small vertebrates[30,49–51] and microbes.[52,53] Here possible applications related to sources of mad cow disease, vectors associated with bird flu, testing compliance

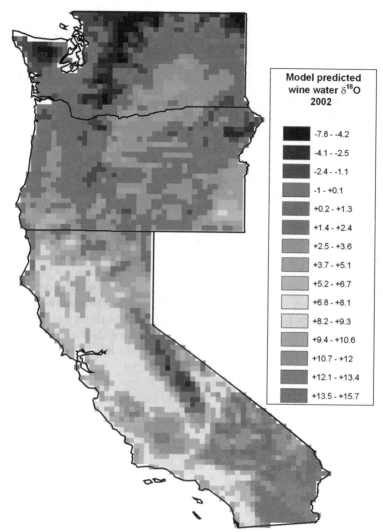

Figure 3   A GIS-based map of the predicted oxygen isotope ratio variations of water in
wine expected on the basis of geographical variations in the isotope ratio of
source water and of the climatic conditions. Based on information published
in West *et al.*[43]

with food production regulations, analyses of CITES-related specimens and
verification of region-of-origin claims, especially for high-value products, are
all contributing to an increased interest in stable isotope analysis in forensic
analyses and food authenticity.[46,54–57]

Hair and fingernails record dietary and water source information and in so
doing provide geographic information for forensic studies. Several recent
studies have suggested that isotope ratio differences in human hair can
be used to distinguish individuals of different geographic origin.[44,46,58,59]

A recent study of public interest involving stable isotopes was the case of the Ice Man discovered on the border between Austria and Italy.[60] Fingernails are also composed of keratin, the same protein as found in hair. Recently, Nardoto et al.[61] have shown that citizens from Europe, USA and Brazil can be distinguished based on the isotope ratios within their fingernails. Although there is a tendency among modern human societies for a global supermarket that would homogenize diets and reduce isotope variations among fingernails of different individuals, the diets in these countries are sufficiently distinct so as still to result in detectable differences in the isotope ratios of human fingernails.

As humans and other mammals move through regions that have water with isotopically distinct values, the δ^2H and $\delta^{18}O$ values of hair should sequentially record that movement. Figure 4 illustrates this point with the isotope ratio analysis of hair from a horse as that animal was raised in Virginia and then transported to Utah. The abrupt changes in the δ^2H and $\delta^{18}O$ values of the hair reflect the transportation of the horse across the USA to its new home. By knowing the growth rate of horse hair, it is possible to reconstruct the temporal patterns of movement. Cerling et al.[30] have developed models to describe the multiple dietary and tissue pools that contribute to determining the turnover rate of amino acids feeding into the synthesis of keratin in hair, providing a mechanistic explanation of why the region-to-region shift in isotope ratios is not instantaneous and also providing dietary insights into the fine-scale

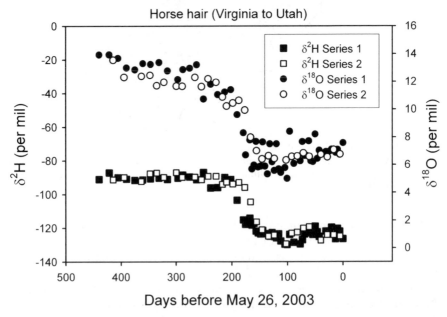

Figure 4 Analysis of the hydrogen and oxygen isotope ratios of sequential tail hair segments from a horse that was moved from Virginia to Utah.

temporal patterns seen in analyses of hair segments. Ehleringer *et al.*[29] have provided a mechanistic model relating the δ^2H and $\delta^{18}O$ values of hair, diet and water source. The inversion of such models allows one to use water isotope maps, such as that in Figure 1, to extrapolate to the possible geographic regions that would be consistent with the geographic movements of the horse, were that information not previously known.

4 Opportunities to Examine C and N Isotopes on a Spatial Basis

Plants and animals often exhibit wide ranges in the carbon isotope ratios of their tissues that are related to geographical patterns.[28,46,62-64] For example, the $\delta^{13}C$ values of human hair can range from –25 to –10‰, reflecting dietary sources in different parts of the world.[59,65,66] This is because carbon isotope ratios in food can also range from –30 to –10‰, depending on the plant's photosynthetic pathway.[28,62,63,67] One challenge is to place carbon isotope ratio variations into a geographical context that is as established in theory as it is for hydrogen and oxygen isotopes described in previous sections.

4.1 The Imprint of Photosynthetic Pathways

Plants can be divided into two primary photosynthetic pathway groups: C$_3$ and C$_4$ photosynthesis.[28] These two pathways exhibit distinctive differences in their $\delta^{13}C$ values, with C$_3$ plants tending to have carbon isotope ratios of about –27‰ and C$_4$ plants about –12‰.[28] Among the foods we eat, most tend to be C$_3$ plants, including most grains, fruits and starchy foods. In contrast, the most common C$_4$ plants are warm-season grasses, which include corn, sorghum, millet and sugarcane.

Figure 5 illustrates the large and non-overlapping differences in the carbon isotope ratios of C$_3$ plants, ranging from –30 to –24‰ (*e.g.* wheat, barley, potatoes), and C$_4$ plants, ranging from –14 to –10‰ (*e.g.* corn, millet, sorghum, sugarcane).[28] It is well established that the isotope ratios of assimilated dietary inputs are subsequently recorded in the proteins, lipids, carbonates and carbohydrates of the muscle, bones, teeth and hair of organisms, providing a record of the diet of that organism.[50,68-76] There are consistent 1–3‰ offsets in the carbon isotope ratios of animal tissues relative to the food substrate that was eaten. Isotopic turnover rates are not constant among body tissues, resulting in the tissue-specific integration of dietary inputs over different temporal periods. Turnover rates of carbon isotopes in blood are on the order of days, whereas turnover rates of bone collagen can be on the order of 4–6 years. Although analyses of these tissues will not yield temporal sequence information related to diet of immediate forensic interest, these measures do provide a longer term record of average dietary inputs. It is only the carbon isotope signal in hair and fingernails that provides a high-resolution temporal record of dietary intake.

Understanding the fine-scale spatial significance of $\delta^{13}C$ values remains a challenge in omnivores, such as humans, because whereas plant-based inputs

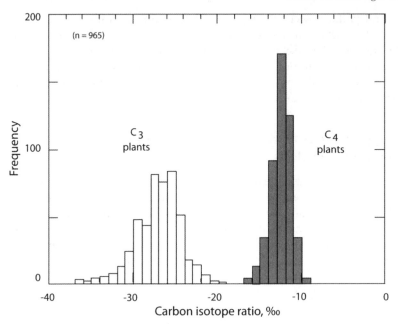

Figure 5 Plants with C_3 versus C_4 photosynthetic pathways have distinct, non-overlapping carbon isotope ratio distributions. Animals that eat C_3 versus C_4
plants will retain an isotopic history of this food source in the hair, bones,
teeth and muscle tissues. Based on Cerling et al.[63]

might reflect local sources, that is not often the case with animal-based protein
sources. For herbivores, the linkages between diet and geography appear to be
more closely related. Several models have been developed that incorporate
spatial models of variations in the $\delta^{13}C$ values of plant communities on a global
scales.[77,78] The Still et al.[78] approach categorized communities as C_3 dominated, C_4 dominated or a mixture of both pathways. Given that the distribution
of C_4 plants is strongly influenced by temperatures during the growing season,
they were able to produce a global $\delta^{13}C$ map based on grid cell temperature
values. Subsequently, the Suits et al.[77] approach is based on C_3 photosynthesis
models with carbon isotope discrimination based on observed meteorology
from the European Center for Medium-Range Weather Forecasts. These approaches have not yet been applied to forensic science, but there are distinct
possibilities of using such approaches to distinguish geographical origins of
border-seized mammals and birds based on $\delta^{13}C$ analyses.

Adulteration of food products through the addition of corn- or sugarcane-
based carbon can be easily observed and is frequently detected in USA foods.
C_4-derived sugars are frequently used to adulterate the carbohydrates contained in jams, jellies and honeys[2,79] and fermentation sources used in making
beer,[80] wines[81] and distilled spirits.[82] However, placing this information on a
geographically based approach is not possible, making it challenging to consider the sourcing or the geographical origins of C_4-adulterated food products.

4.2　Cocaine Origins are Reflected in C and N Isotopes

For two decades, there has been international forensic interest in using stable isotope ratio analyses to analyze drugs of abuse, particularly heroin and cocaine.[83–85] Approximately a decade ago, the US DEA launched an effort to obtain sufficient authentic specimens with which to develop a region-of-analysis program. These results were positive. Dual-isotope plots described different spatial realms, providing an approach for distinguishing regions of interest. As part of this collaborative effort, Ehleringer *et al.*[11] showed that the major growing regions of coca for cocaine in South America could be characterized by plots of the carbon versus nitrogen isotope ratios of purified cocaine (Figure 6). In this case, the combination of these two isotopes alone explained more than 80% of the observed variation among samples originating from different growing regions. This information is useful in law enforcement for strategic and intelligence purposes. Today, stable isotope analyses are a key measurement in the DEA's Cocaine Signature Program, providing critical information about the origins of seized samples.

The challenge is to translate the information contained in the carbon and nitrogen isotope ratios of cocaine into a semi-mechanistic model that can be integrated into GIS mapping, as was shown earlier for hydrogen and oxygen isotopes. Because the mixing of carbon dioxide into different layers of the atmosphere is relatively rapid, there are no pronounced spatial gradients in the isotope ratios of atmospheric carbon dioxide, the photosynthetic substrate that forms the basis of all carbon in plants. Instead, there are variations in the fractionation against ^{13}C during photosynthesis.[28] These fractionation events are driven by the degree of stomatal opening,[86–88] leading to regional-to-continental patterns associated with different climate zones.[89–91] Thus, carbon isotope

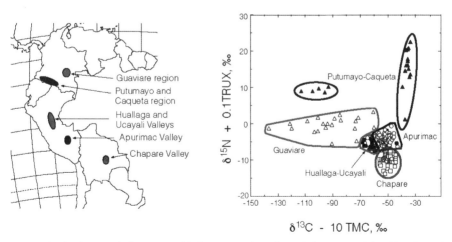

Figure 6　Carbon and nitrogen stable isotope ratios allowed the correct identification of country of origin of 90% of 200 + cocaine samples analyzed (from Bolivia, Colombia or Peru). Based on data in Ehleringer *et al.*[11]

variations among C_3 plants are likely to follow precipitation clines, especially since humidity and precipitation are so closely co-linked across sites. Although the challenge of placing C_3-based ^{13}C variations on to a global scale is challenging, the coherent patterns that appear in the carbon-nitrogen isotope plot of Figure 6 provide support for potential forensic applications once these modeling challenges have been met.

References

1. C. Brunnee, *Rapid Commun. Mass Spectrom.*, 1997, **11**, 694–707.
2. J. R. Ehleringer, T. E. Cerling and J. B. West, in *Forensic Analysis on the Cutting Edge: New Methods for Trace Evidence Analysis*, ed. R. D. Blackledge, Wiley, San Diego, 2007, pp. 399–422.
3. J. B. West, G. J. Bowen, T. E. Cerling and J. R. Ehleringer, *Trends Ecol. Evolut.*, 2006, **21**, 408–414.
4. G. J. Bowen, L. I. Wassenaar and K. A. Hobson, *Oecologia*, 2005, **143**, 337–348.
5. S. Benson, C. Lennard, P. Maynard and C. Roux, *Forensic Sci. Int.*, 2006, **157**, 1–22.
6. W. Meier-Augenstein and R. H. Liu, in *Advances in Forensic Applications of Mass Spectrometry*, ed. J. Yinon, CRC Press, Boca Raton, FL, 2004, pp. 149–180.
7. J. T. Brenna, T. N. Corso, H. J. Tobias and R. J. Caimi, *Mass Spectrom. Rev.*, 1997, **16**, 227–258.
8. K. Pye and D. J. Croft, eds. *Forensic Geoscience: Principles, Techniques and Applications*, Geological Society, London, 2004.
9. B. L. Beard and C. M. Johnson, *J. Forensic Sci.*, 2000, **45**, 1049–1061.
10. J. R. Ehleringer, D. A. Cooper, M. J. Lott and C. S. Cook, *Forensic Sci. Int.*, 1999, **106**, 27–35.
11. J. R. Ehleringer, J. F. Casale, M. J. Lott and V. L. Ford, *Nature*, 2000, **408**, 311–312.
12. J. Casale, E. Casale, M. Collins, D. Morello, S. Cathapermal and S. Panicker, *J. Forensic Sci.*, 2006, **51**, 603–606.
13. E. K. Shibuya, J. E. S. Sarkis, O. Negrini and L. A. Martinelli, *Forensic Sci. Int.*, 2006, **167**, 8–15.
14. T. M. Denton, S. Schmidt, C. Critchley and G. R. Stewart, *Aust. J. Plant Physiol.*, 2001, **28**, 1005–1012.
15. P. W. Rundel, J. R. Ehleringer and K. A. Nagy, eds. *Stable Isotopes in Ecological Research*, Springer, New York, 1988.
16. H. Griffiths, ed. *Stable Isotopes Integration of Biological, Ecological and Geochemical Processes*, BIOS Scientific Publishers, Oxford, 1998.
17. J. R. Ehleringer, A. E. Hall and G. D. Farquhar, eds. *Stable Isotopes and Plant Carbon/Water Relations*, Academic Press, San Diego, 1993.
18. A. Dutton, B. H. Wilkinson, J. M. Welker, G. J. Bowen and K. C. Lohmann, *Hydrol. Process.*, 2005, **19**, 4121–4146.
19. G. Bowen, J. Ehleringer, L. Chesson, E. Stange and T. Cerling, *Water Resour. Res.*, 2007, **43**, W03419.

20. J. R. Gat, *Annu. Rev. Earth Planet. Sci.*, 1996, **24**, 225–262.
21. K. Rozanski, L. Araguas-Araguas and R. Gonfiantini, *Science*, 1992, **258**, 981–985.
22. Y. Yurtsever, *Worldwide Survey of Stable Isotopes in Precipitation*, Report of the Section of Isotope Hydrology, IAEA, Vienna, 1975.
23. G. J. Bowen and J. Revenaugh, *Water Resour. Res* 2003, **39**, 1299.
24. G. J. Bowen and B. Wilkinson, *Geology*, 2002, 30.
25. G. J. Bowen, D. A. Winter, H. J. Spero, R. A. Zierenberg, M. D. Reeder, T. E. Cerling and J. R. Ehleringer, *Rapid Commun. Mass Spectrom.*, 2005, **19**, 3442–3450.
26. G. J. Bowen, T. E. Cerling and J. R. Ehleringer, in *Stable Isotopes as Indicators of Ecological Change*, eds. T. E. Dawson and R. T. W. Siegwolf, Academic Press, San Diego, 2007, pp. 285–300.
27. G. D. Farquhar, J. Lloyd, J. A. Taylor, L. B. Flanagan, J. P. Syvertsen, K. T. Hubick, S. C. Wong and J. R. Ehleringer, *Nature*, 1993, **363**, 439–443.
28. G. D. Farquhar, J. R. Ehleringer and K. T. Hubick, *Annu. Rev. Plant Physiol. Plant Mol. Biol.*, 1989, **40**, 503–537.
29. J. R. Ehleringer, G. J. Bowen, L. A. Chesson, A. G. West, D. Podlesak and T. E. Cerling, *Proc. Natl. Acad. Sci. USA*, in press.
30. T. E. Cerling, L. K. Ayliffe, M. D. Dearing, J. R. Ehleringer, B. H. Passey, D. W. Podlesak, A. M. Torregrossa and A. G. West, *Oecologia*, 2007, **151**, 175–189.
31. J. S. Roden, G. G. Lin and J. R. Ehleringer, *Geochim. Cosmochim. Acta*, 2000, **64**, 21–35.
32. J. S. Roden and J. R. Ehleringer, *Plant Physiol.*, 1999, **120**, 1165–1173.
33. J. S. Roden and J. R. Ehleringer, *Oecologia*, 2000, **123**, 481–489.
34. L. Sternberg and M. J. DeNiro, *Science*, 1983, **220**, 947–949.
35. C. J. Yapp and S. Epstein, *Geochim. Cosmochim. Acta*, 1982, **46**, 955–965.
36. J. B. West, A. Sobek and J. R. Ehleringer, *PLoS Biology*, 2007, in review.
37. I. Silverman, *New Yorker*, October 23, 1995, 50.
38. S. Mihm, *New York Times*, July 23, 2006.
39. N. L. Ingraham and E. A. Caldwell, *J. Geophys. Res.*, 1999, **102**, 2185–2194.
40. I. J. Kosir, M. Kocjaneic, N. Ogrinc and J. Kidric, *Anal. Chim. Acta*, 2001, **429**, 195–206.
41. G. Martin, D. Odiot, V. Godineau and N. Naulet, *Appl. Geochem.*, 1989, **4**, 1–11.
42. G. Martin, C. Guillou, M. L. Martin, M. T. Cabanis, Y. Tep and J. Aerny, *J. Agric. Food Chem.*, 1988, **36**, 316–322.
43. J. B. West, J. R. Ehleringer and T. E. Cerling, *J. Agric. Food Chem.*, 2007, **55**, 7075–7083.
44. I. Fraser, W. Meier-Augenstein and R. M. Kalin, *Rapid Commun. Mass Spectrom.*, 2006, **20**, 1109–1116.
45. D. M. O'Brien and M. J. Wooler, *Rapid Commun. Mass Spectrom.*, 2007, **21**, 2422–2430.

46. Z. D. Sharp, V. Atudorei, H. O. Panarello, J. Fernandez and C. Douthitt, *J. Archaeol. Sci.*, 2003, **30**, 1709–1716.

47. K. A. Hobson and L. I. Wassenaar, *Oecologia*, 1997, **109**, 142–148.

48. C. P. Chamberlain, J. D. Blum, R. T. Holmes, F. Xiahong, T. W. Sherry and G. R. Graves, *Oecologia*, 1997, **109**, 132–141.

49. K. A. Hobson, L. Atwell and L. I. Wassenaar, *Proc. Natl. Acad. Sci. USA*, 1999, **96**, 8003–8006.

50. B. H. Passey, T. F. Robinson, L. K. Ayliffe, T. E. Cerling, M. Sponheimer, M. D. Dearing, B. L. Roeder and J. R. Ehleringer, *J. Archaeol. Sci.*, 2005, **32**, 1459–1470.

51. D. W. Podlesak, A.-M. Torregrossa, J. R. Ehleringer, M. D. Dearing, B. H. Passey and T. E. Cerling, *Geochim. Cosmochim. Acta.*, 2008, **72**, 19–35.

52. H. W. Kreuzer-Martin and K. H. Jarman, *Appl. Environ. Microbiol.*, 2007, **73**, 3896–3908.

53. H. W. Kreuzer-Martin, M. J. Lott, J. Dorigan and J. R. Ehleringer, *Proc. Natl. Acad. Sci. USA*, 2003, **100**, 815–819.

54. M. Boner and H. Forstel, *Anal. Bioanal. Chem.*, 2004, **378**, 301–310.

55. S. Kelly, K. Heaton and J. Hoogewerff, *Trends Food Sci. Technol.*, 2005, **16**, 555–567.

56. A. Rossmann, *Food Rev. Int.*, 2001, **17**, 347–381.

57. L. A. Chesson, A. H. Thompsaon, D. W. Podlesak, T. E. Cerling and J. R. Ehleringer, *Journal of Agricultural and Food Chemistry*, 2007, in review.

58. T. C. O'Connell, R. E. M. Hedges, M. A. Healey and A. H. R. Simpson, *J. Archaeol. Sci.*, 2001, **28**, 1247–1255.

59. T. C. O'Connell and R. E. M. Hedges, *Am. J. Phys. Anthropol.*, 1999, **108**, 409–425.

60. S. A. Macko, G. Lubec, M. Teschler-Nicola, V. Andrusevich and M. H. Engel, *FASEB J.*, 1999, **13**, 559–562.

61. G. Nardoto, S. Silva, C. Kendall, J. Ehleringer, L. Chesson, E. Ferraz, M. Moreira, J. Ometto and L. Martinelli, *Am. J. Phys. Anthropol.*, 2006, **131**, 137–146.

62. J. R. Ehleringer, T. E. Cerling and B. R. Helliker, *Oecologia*, 1997, **112**, 285–299.

63. T. E. Cerling, J. M. Harris, B. J. MacFadden, M. G. Leakey, J. Quade, V. Eisenmann and J. R. Ehleringer, *Nature*, 1997, **389**, 153–158.

64. R. Michener and K. Lajtha, eds. *Stable Isotopes in Ecology and Environmental Science*, 2nd edn., Blackwell, Malden, MA, 2007.

65. R. Bol, J. Marsh and T. H. E. Heaton, *Rapid Commun. Mass Spectrom.*, 2007, **21**, 2951–2954.

66. R. Bol and C. Pflieger, *Rapid Commun. Mass Spectrom.*, 2002, **16**, 2195–2200.

67. T. E. Cerling, G. Wittemyer, H. B. Rasmussen, F. Vollrath, C. E. Cerling, T. J. Robinson and I. Douglas-Hamilton, *Proc. Natl. Acad. Sci. USA*, 2006, **103**, 371–373.

68. S. Jim, S. H. Ambrose and R. P. Evershed, *Geochim. Cosmochim. Acta*, 2004, **68**, 61–72.

69. S. H. Ambrose, *J. Hum. Evol.*, 1986, **15**, 707–731.

70. T. F. Robinson, M. Sponheimer, B. L. Roeder, B. Passey, T. E. Cerling, M. D. Dearing and J. R. Ehleringer, *Small Ruminant Res.*, 2006, **64**, 162–168.
71. M. Sponheimer, T. Robinson, L. K. Ayliffe, B. H. Passey, B. Roeder, L. Shipley, E. Lopez, T. E. Cerling, D. Dearing and J. Ehleringer, *Can. J. Zool. Rev. Can. Zool.*, 2003, **81**, 871–876.
72. L. K. Ayliffe, T. E. Cerling, T. Robinson, A. G. West, M. Sponheimer, B. H. Passey, J. Hammer, B. Roeder, M. D. Dearing and J. R. Ehleringer, *Oecologia*, 2004, **139**, 11–22.
73. K. A. Hobson and R. G. Clark, *Auk*, 1993, **110**, 638–641.
74. K. A. Hobson, R. T. Alisauskas and R. G. Clark, *Condor*, 1993, **95**, 388–394.
75. M. J. DeNiro and S. Epstein, *Science*, 1977, **197**, 261–263.
76. M. J. DeNiro and S. Epstein, *Geochim. Cosmochim. Acta*, 1978, **42**, 495–506.
77. N. S. Suits, A. S. Denning, J. A. Berry, C. J. Still, J. Kaduk J. B. and Miller I. T. Baker, *Global Biogeochem. Cycles*, 2005, **19**. doi: 10.1029/2003GB002141.
78. C. J. Still, J. A. Berry G. J Collatz and R. S. DeFries, *Global Biochem. Cycles*, 2003 **17**. doi: 10.1029/2001GB01807.
79. J. W. White, K. Winters, P. Martin and A. Rossman, *J. AOAC Int.*, 1998, **81**, 610–619.
80. J. R. Brooks, N. Buchmann, S. Phillips, B. Ehleringer, R. D. Evans, M. Lott, L. A. Martinelli, W. T. Pockman, D. R. Sandquist, J. P. Sparks, L. Sperry, D. Williams and J. R. Ehleringer, *J. Agric. Food Chem.*, 2002, **50**, 6413–6418.
81. L. A. Martinelli, M. Z. Moreira, J. P. Ometto, A. R. Alcarde, L. A. Rizzon, E. Stange and J. R. Ehleringer, *J. Agric. Food Chem.*, 2003, **51**, 2625–2631.
82. L. Pissinatto, L. A. Martinelli, R. L. Victoria and P. B. de Camargo, *Food Res. Int.*, 2000, **32**, 665–668.
83. F. Besacier, R. Guilluy, J. L. Brazier, H. Chaudron-Thozet, J. Girard and A. Lamotte, *J. Forensic. Sci.* 1997, **42**, 429–433.
84. M. Desage, R. Guilluy and J. L. Brazier, *Analytica Chimica. Acta.*, 1991, **247**, 249–254.
85. F. Besacier, H. Chaudron-Thozet, M. Rousseau-Tsangaris, J. Girard and A. Lamotte, *Forensic Sci. Intl.*, 1997, **85**, 113–125.
86. J. R. Ehleringer and T. E. Cerling, *Tree Physiol.*, 1995, **15**, 105–111.
87. J. R. Ehleringer, in *Ecophysiology of Photosynthesis*, Ecological Studies Seriesed. eds. E.-D. Schulze and M. M. Caldwell, Springer, New York, 1994, pp. 361–392.
88. J. R. Ehleringer, in *Water Deficits: Plant Responses From Cell to Community*, eds. S. J. A. C. and H. Griffiths, BIOS Scientific, London, 1993, pp. 265–284.
89. J. M. Miller, R. J. WIlliams and G. D. Farquhar, *Funct. Ecol.*, 2001, **15**, 222–232.
90. J. P. Comstock and J. R. Ehleringer, *Proc. Natl. Acad. Sci. USA*, 1992, **89**, 7747–7751.
91. G. R. Stewart, M. H. Turnbull, S. Schmidt and P. D. Erskine, *Aust. J. Plant Physiol.*, 1995, **22**, 51–55.

Diagnostic Compounds for Fingerprinting Petroleum in the Environment

SCOTT A. STOUT AND ZHENDI WANG

1 Introduction

The term 'petroleum' refers to a family of gaseous, liquid and solid hydro-carbon-rich materials that are both naturally occurring (crude) and man-made (refined). Crude petroleum includes natural gas and gas condensate, light and heavy crude oil, oil shale/sand and solid bitumen; refined petroleum includes many different fuels, solvents, petrochemicals, lubricants, waxes, asphalt and other products. Because economical alternatives to most petroleum products are rare or slow in coming, the widespread use of petroleum over many decades has led to its widespread occurrence in the environment – with sources ranging from large marine oil spills to leaking underground storage tanks to the in-complete combustion of fuels. Establishing the source(s) of petroleum-derived contamination is often at the center of environmental forensic investigations.[1,2] The complex chemical makeup of petroleum – which can contain tens of thousands of individual hydrocarbons and non-hydrocarbons – provides an opportunity to 'chemically fingerprint' petroleum contamination and thereby assess its relationship to known or suspected sources.

Significant advances in the chemical fingerprinting of petroleum contamin-ation for environmental forensic purposes have occurred in the past 15 years. Many of the advances have followed advances in petroleum geochemistry – *i.e.* the chemistry applied to the exploration and production of crude petroleum – wherein the long-recognized diagnostic features of crude petroleum have been applied in environmental studies. For example, petroleum biomarkers – chemicals long utilized by petroleum exploration geochemists[3] – have found widespread use in forensic investigations.[4] Other advances are the result of the improved understanding of the effects of environmental processes (*e.g.* weathering and mixing) on the chemistry of fugitive petroleum.[5,6] Still other advances are driven by the development or application of new analytical instrumentation.[7–9]

Issues in Environmental Science and Technology, No. 26
Environmental Forensics
Edited by RE Hester and RM Harrison
© Royal Society of Chemistry 2008

Table 1 Inventory of diagnostic compound groups discussed.

Diagnostic compound group	Approximate carbon boiling range	Mass spectral fragment ions (m/z)
Trimethylpentanes	C_7 to C_8	57, 71
Lead alkyls	C_7 to C_{12}	223
Alcohols and ethers	C_5 to C_7	Various
n-Alkanes	C_1 to C_{45}[a]	85
Acyclic isoprenoids	C_{12} to C_{19}	113
n-Alkylcyclohexanes	C_6 to C_{45}[a]	83
Adamantanes and diamantanes	C_{10} to C_{15}	Various
Bicyclic sesquiterpanes	C_{13} to C_{17}	123
Diterpanes	C_{19} to C_{24}	191
Polycyclic aromatic hydrocarbons	C_{10} to C_{45}[a]	Various
Extended tricyclic triterpanes	C_{18} to C_{26}	191
Tetracyclic terpanes	C_{22} to C_{23}	191
25-Norhopanes (10-desmethylhopanes)	C_{25} to C_{33}	177
Hopanes and other pentacyclic triterpanes	C_{26} to C_{34}	191
Diasteranes	C_{23} to C_{26}	217
Regular steranes	C_{25} to C_{29}	217 and 218
C-ring monoaromatic steroids	C_{23} to C_{26}	253
A,B,C-ring triaromatic steranes	C_{26} to C_{29}	231

[a]As measured by conventional (low-temperature) gas chromatography.

In this chapter, it is our objective to review some of the most useful diagnostic compounds or compound groups (Table 1) that have been used in the chemical fingerprinting of petroleum-derived contamination. The compound groups described herein span the full boiling range of most petroleum and include both naturally occurring and man-made (refined additives) chemicals. The remainder of this section presents some background information on the factors affecting the chemical fingerprint of petroleum in the environment (Section 1.1) and, because chemical fingerprinting can mean different things to different scientists, we also offer some philosophy on its meaning with respect to petroleum in environmental forensic investigations (Section 1.2).

1.1 Petroleum Genesis, Refining, Weathering and Mixing

The 'chemical fingerprint' of petroleum or petroleum-impacted matrices in the environment will be determined by four predominant factors: (1) the conditions of the crude petroleum's genesis over geological time, (2) the effects of any refining processes, (3) the extent of environmental weathering it experiences and (4) the degree to which it may be mixed with unrelated contamination or naturally occurring hydrocarbons after being released into the environment. The basis for each of these four factors is discussed in the following paragraphs, with more detail available elsewhere.[6]

The chemistry of naturally occurring petroleum – gases, liquids and solids – is the collective result of the geological processes that led to their formation over millions of years, *viz.* (1) the nature of the ancient, organic-rich, source strata, (2) the thermal history of the source strata and (3) changes brought about during petroleum's migration to and residence in reservoir strata.[10] These three processes collectively impart very specific chemical features to naturally occurring petroleum, which vary enormously worldwide and even within multiple reservoirs within a single oil field.[11] These *genetic* compositional features provide a strong basis for the chemical fingerprinting of naturally occurring petroleum after its release into the environment.[1]

Man-made petroleum products are derivatives of naturally occurring petroleum produced in the course of petroleum refining – including both the authentic products (*e.g.* fuels and lubricants) and their derivatives (*e.g.* uncombusted fuel emissions and waste oils). Although some of the genetic chemical features of the *parent* crude petroleum are passed directly to the *daughter* petroleum products and wastes during refining, the refining process can impart its own effects on the chemical composition of man-made petroleum.[6,12] The effects of refining are perhaps most evident in automotive gasoline, which nowadays is largely comprised of man-made blending components produced during refining, *e.g.* isomerate, reformate, alkylate and FCC gasoline.[13] Other petroleum products include 'cracked' intermediates or additives that are absent from naturally occurring petroleum. Thus, in total, the chemical compositions and fingerprints of refined petroleum products result from the combined effects of the parent crude petroleum and the refining process.

Once crude or refined petroleum is released into the environment, it is immediately subject to transformation due to biotic (aerobic and anaerobic biodegradation) and abiotic (dissolution, evaporation, sorption) processes that are collectively termed 'weathering'. Weathering will alter the chemical composition or fingerprinting of petroleum in the environment, often at widely differing rates. The rate of weathering of petroleum depends on numerous site-specific factors that will vary by many scales. Among these factors, clearly the nature of the petroleum itself is important because certain types are more susceptible to weathering than others (*e.g.* gasoline *versus* lubricating oil). The local and microscale conditions (*e.g.* oxygen availability, temperature and chemical concentrations) will also affect the rate of weathering. Collectively, it is the combined effects of the *rate* of weathering (controlled by the aforementioned factors) and the *duration* of a petroleum's tenure in the environment that will establish the overall **extent** of weathering, that is,

$$\text{rate} \times \text{time} = \text{extent}$$

Hence time alone does not determine the extent of weathering of petroleum in the environment, despite empirical-based claims to the contrary.[14] Since site-specific factors can change over time, naturally the rate of weathering can change over time. However, an important fact with respect to chemical fingerprinting is that regardless of the rate, the weathering of petroleum in the environment proceeds in a fairly predictable and progressive manner, *i.e.* (1) some compounds are more susceptible to weathering than others and (2) weathering is not a reversible process.[15]

Concurrent with weathering, once petroleum is released into the environment it is subject to mixing with pre-existing anthropogenic contaminants or naturally occurring materials already present in the environment. These 'background' hydrocarbons can contribute to the collective fingerprint of any samples thought to be impacted by a spilled petroleum only.[16] If unrecognized, the presence of pre-existing, background hydrocarbons can confound chemical fingerprinting studies that rely upon comparisons of suspected sources with suspected impacted samples.

1.2 The Philosophy of Chemical Fingerprinting

Although sometimes the objective of chemical fingerprinting is simply to characterize an unknown petroleum, the objective for most environmental forensic investigations involving petroleum contamination is to determine the relationship between suspected source(s), impacted matrices and/or background samples. Assessing any such relationship is essentially a correlation exercise in which the chemical features of one sample are compared with those of another. Comparisons should preferentially rely upon 'genetic' (*i.e.* source-dependent and weathering-independent) parameters and necessarily consider the aforementioned effects of mixing. In the past, such comparisons most often were conducted qualitatively, in which a fingerprint for one sample might be visually compared with that of another, sometimes using a light table. Some oil spill identification protocols still rely upon such qualitative comparisons,[17,18] despite the subjective nature of such correlations. Modern analytical techniques in which chromatographic or spectral peak areas or heights are measured allows for semiquantitative (*e.g.* peak ratios) or, through the use of internal standards and calibrations, fully quantitative (*e.g.* absolute concentrations) comparisons to be made. Either of these provides sample-specific metrics that are amenable to statistical or numerical correlation or pattern matching techniques, collectively called 'chemometrics', that have been more recently employed in oil spill identification protocols.[19–22] These chemometric correlations can involve simple binary or tertiary plots of semi- or fully-quantitative of two or three metrics or more sophisticated statistical or multivariate analyses of many metrics. Either approach can serve to demonstrate the degree of similarity between multiple samples being compared. Although we favor the use of quantitative correlation tools in petroleum fingerprinting, which imparts a certain degree of impartiality, we maintain that qualitative visual examination of chemical fingerprints cannot be ignored since it can reveal problems or otherwise unnoticed details that may not be evident among the selected metrics. The human eye remains a powerful chemical fingerprinting tool that should be used in combination with chemometric tools.

As will be evident in Section 2, the specific types of chemicals that are diagnostic among different sources of petroleum are not those commonly measured using standardized EPA or other analytical methods. This is not surprising since the standard methods were not developed for environmental forensic purposes but were instead intended to define the nature and extent of contamination.[23] These standard methods can, however, be modified so that chemical fingerprinting objectives can be achieved often simply by targeting diagnostic compounds rather than the prescribed Priority Pollutant target compounds.[24] Chemical fingerprinting of petroleum contamination is hampered when the available data are limited to those hydrocarbons or other compounds that can be universally present in petroleum of any type [e.g. BTEX or Priority Pollutant polycyclic aromatic hydrocarbons (PAHs)]. Thus, environmental forensic investigations involving petroleum-impacted samples and their suspected sources require more sophisticated and source-specific data than are necessary for regulatory (nature and extent or risk) purposes.

It is fair to question what constitutes a 'positive' or 'negative' chemical correlation between petroleum or petroleum-impacted samples. This is a difficult question to answer and certainly is beyond the scope of this chapter. The reader is directed to previous publications that discuss the use and application of multiple similarity indices or correlation tools (e.g. various types of correlation coefficients, cosine theta, Student's t-test, analysis of variance, confidence limits, null hypothesis testing).[19,22,25–28] Our point to be made here is that any attempt at correlation requires that the analytical precision of the data used be robustly established through the analysis of replicate samples and reference materials. Correlations are empirical in the sense that their results are only as good as the input data that are used. The importance of analytical precision is obvious when one considers that different levels of precision (e.g. as might be achieved at different laboratories) could well yield different correlations – most often with a false 'positive' correlation achieved when less precise data (with a larger statistical error than for more precise data) are used.

Generally, it should be evident that the more diagnostic metrics are used, the stronger a case can be made for a positive correlation – so long as the diagnostic metrics are both resistant to weathering, precisely measurable and variable among different petroleum sources (see Section 2.3). For example, a positive correlation based on a small number of chemical features, which might fortuitously match in unrelated petroleums, is subject to careful scrutiny. It is for this reason that modern protocols for oil spill correlations rely upon a large number of diagnostic parameters. Some diagnostic parameters might be more meaningful and source-specific than others, e.g. the prominence of an unusual biomarker such as gammacerane (see Section 2.9) might invoke greater confidence in a positive correlation than a pristane/phytane ratio. Positive chemical correlations between samples usually, but not necessarily, imply a genetic relationship between the samples. Specifically, a positive chemical correlation should also be reconciled with the geological or hydrological circumstances and the available facts or historical circumstances. It is because of this need for some form of 'ground-truthing' that chemical fingerprinting constitutes only one component – albeit an important one – of most environmental forensic investigations.

Negative chemical correlations, on the other hand, require identification of only one chemical difference that cannot be explained by natural variation among a related population of oils (e.g. inhomogeneities within a subsurface oil pool), variations attributable to matrix effects (e.g. partitioning effects when comparing oil with soil, water or tissue), the analytical precision of the measurement or by mixing or weathering in the environment. Because of these possibilities, even negative correlations require careful scrutiny before a conclusion is drawn as to the lack of a relationship between two petroleum-impacted samples. If the potential for mixing is significant (*e.g.* mixing with background petroleum at low levels or co-mingling of multiple petroleum sources), then additional numerical models that address allocation or source apportionment[29] should be considered before a negative correlation is concluded. The difficulty with source allocation models for mixed samples is that they will rarely result in a 0% contribution from any given potential source. Hence mathematical solutions or allocations require careful scrutiny as to the uniqueness of their solution(s). It should be noted that fully quantitative data (absolute concentrations) are considered necessary for any quantitative source apportionment.[30]

Any chemical fingerprinting investigation of petroleum must also ask if the sample suite is sufficiently large as to represent any natural variation in the nature of the contamination or suspected source material(s). Perhaps the most common situation where the natural variation must be considered is in large masses of free phase petroleum in the subsurface that often exhibit spatial heterogeneities within the pool, often arising from incomplete mixing of chemically disparate petroleum released over a long period. In such instances, the petroleum from a single monitoring well would inadequately characterize the contamination and perhaps lead to a false negative correlation.

Our necessarily brief discussion of the philosophy of chemical fingerprinting of petroleum contamination should emphasize that we consider (1) the combination of qualitative and quantitative fingerprinting techniques, (2) the need for precise analytical data for petroleum-specific analytes beyond those conventionally acquired for regulatory purposes and (3) careful and unbiased application of a statistical or numerical correlation tool(s) as important components of any chemical fingerprinting investigation.

The discussion above has implied that chemical fingerprinting of petroleum necessarily relies upon a large number of individual diagnostic compounds or compound groups with known identities be measured and compared. Although this is the most common approach to chemical fingerprinting, one must acknowledge that the specific identities of individual compounds might not be necessary when comparing chromatographic or spectrographic patterns. Arguments in favor of this approach include the ability to include all or nearly all of the data that have been collected, instead of only selected diagnostic compounds that represent only a very small fraction of an oil's total mass and that are often chosen *a priori*. For example, depending on the run time and scan range, a single gas chromatographic/mass spectrometric (GC–MS) analysis can produce millions of data points that are available for numerical analysis. Recent workers have attempted to use pattern recognition software to correlate digitized chromatographic data for petroleum-impacted environmental samples.[20,31] Although this approach may find wider application in the future, most environmental forensic investigations involving petroleum will continue to rely on the use of diagnostic compounds and ratios, upon which we have focused here.

2   Diagnostic Compounds

There is no prescribed list of diagnostic compounds that is universally appropriate for the chemical fingerprinting of petroleum. Although much use has been made of the now familiar sterane and triterpane biomarkers (Sections 2.9 and 2.10[32]), the relatively high boiling range of these compounds (Table 1) renders them inappropriate for studies involving lower boiling petroleum fractions or fuels in which their concentrations are extremely low or nil. In other instances, even the list of diagnostic compounds may need to evolve once some preliminary analyses are conducted to determine what exactly *is* diagnostic for the oil(s) in question.

Generally, the characteristics of any diagnostic compounds should include (1) the potential for a high degree of specificity among different oils or fuels, (2) relative widespread occurrence in petroleum, (3) ability to be precisely measured within a complex mixture (*e.g.* co-elution issues can be overcome) and in multiple matrices (*e.g.* oil, sludge, soil, sediment, tissues, water, vapor, air) and, perhaps most importantly, (4) relative resistance to environmental weathering. Contrary to (2) above, in some instances a 'unique' compound that is not necessarily widespread may prove extremely diagnostic. The distribution of diagnostic compounds or ratios among them [A/B or A/(A + B)] are often used to represent the character of a given set of samples. When ratios are used they should generally involve homologues, stereoisomers or simply compounds with comparable chemical structures and/or chemico-physical properties, which minimize the effect of environmental weathering or partitioning on the ratio.

In this chapter, we focus on those diagnostic compounds amenable to study by GC–MS, since this analytical method has been adopted by most environmental laboratories conducting chemical fingerprinting investigations. Modern benchtop single quadrupole GC–MS systems offer sufficient capabilities for most environmental fingerprinting applications, although larger, more expensive floor models will offer added capabilities such as tandem mass spectrometry, metastable reaction monitoring or compound-specific isotope ratio analysis. GC–MS is extremely useful in the study of petroleum since it permits detailed separations and identifications of individual compounds that may be present in only minute quantities in petroleum and petroleum-impacted samples. Table 1 lists the diagnostic compounds or compound groups discussed here and provides some general information on their mass spectral analysis and carbon boiling ranges.

2.1   Trimethylpentanes

Although present in minute concentrations in crude petroleum, the four isomers of trimethylpentane (2,2,4-, 2,2,3-, 2,3,4- and 2,3,3-TMP) are the primary products of the refining process known as alkylation.[33] The product of the alkylation process, or alkylate, is commonly used in blending automotive and aviation gasolines because of its high octane and volatility. Two acid-catalyzed processes – with H_2SO_4 and HF – are used and these have been demonstrated to produce the TMP isomers in varying proportions. Specifically, the proportion of isooctane (2,2,4-TMP) is generally greater in alkylate produced by the HF method regardless of the olefin feed.[34–36] The measurement of the TMP isomers in fuels and environmental matrices can be achieved using GC–MS in combination with authentic standards.[24] The utility of the variation in TMP isomer patterns in forensic investigations of gasolines was first discussed by Beall *et al.*,[37] who used published TMP yield data to calculate the percentage of isooctane relative to all four TMP isomers using the following formula:

$$2,2,4\text{-}TMP/(2,2,4\text{-}TMP + 2,2,3\text{-}TMP + 2,3,4\text{-}TMP + 2,3,3\text{-}TMP)$$

The results indicated that gasolines blended using an HF alkylate would be expected to contain a higher percentage of isooctane (~ 54–73%) than gasolines blended using an H_2SO_4 alkylate (~ 39–45%), a result consistent with the early literature (above) and more recently supported by TMP concentrations measured in gasolines containing alkylate of known origins (Figure 1[38]). Since the TMP isomers would be expected, based on their similar physico-chemical properties,[39] to react similarly to environmental weathering, the percentage of isooctane (of the total TMP) can be diagnostic in distinguishing automotive and aviation gasolines blended using a HF-type versus H_2SO_4-type alkylates.

2.2 Gasoline Additives

Automotive gasoline has historically contained additives that are intended to (1) increase power/reduce antiknock (*i.e.* increase octane), (2) remove or prevent carburetor and spark plug deposits, (3) prevent the formation of gums, peroxides, rust and ice in engines, (4) decrease tailpipe emissions and (5) distinguish different products or gasoline grades.[13] Two of the more significant additives blended into gasoline, both which have forensic applications, are lead alkyls and oxygenates, *i.e.* the family of oxygen-containing compounds including both alcohols (*e.g.* ethanol) and ethers [*e.g.* methyl *tert*-butyl ether (MTBE)],[40] upon which we have focused here.

2.2.1 Lead Alkyls. Prior to being banned in most countries over the past 10 years or so, lead alkyls were common octane enhancers found in automotive gasoline.[41] The measurement of these compounds can be used to determine the total lead concentration in a fugitive gasoline, potentially age-constraining its vintage[38,41–43] or deduce the source of the gasoline based on the distribution of the five individual lead alkyls historically used.[42] The basis for the latter involves the predictable distribution of the five individual lead alkyls — tetramethyllead (TML; m/z 253, 223), trimethylethyllead (TMEL; m/z 253, 223), dimethyldiethyllead (DMDL; m/z 267, 223), methyltriethyllead (MTEL; m/z 281, 223) and tetraethyllead (TEL; m/z 295, 237) — that were contained in the limited number of antiknock additive packages available for use in different gasolines. Refiners often blended specific antiknock additive packages (Table 2)

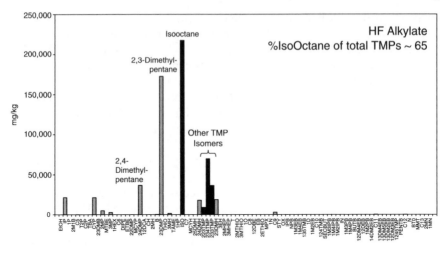

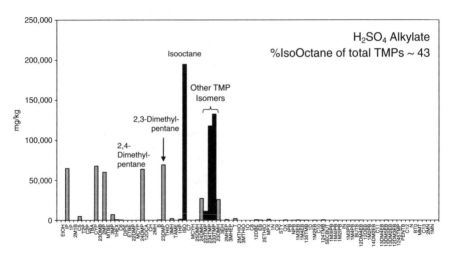

Figure 1   Histograms showing the concentrations (mg kg_{oil}^{-1}) of volatile compounds contained in alkylate (gasoline blend stock) produced using the (A) hydrofluoric acid (HF)-catalyzed alkylation process and (B) sulfuric acid (H_2SO_4) alkylation process. The prominent isoalkanes include the trimethylpentane (TMP) isomers discussed in the text. Abbreviations of other target compounds are from Douglas et al. (2007).[24]

in different grades or markets that, when reconciled with historic blending information, might prove diagnostic to a specific source.

The five individual alkyl lead compounds can be measured by GC–MS in the selected ion monitoring mode following adaptations of EPA Method 8270 with detection limits as low as $5\,\mu g\,ml^{-1}$ in non-aqueous phase liquids (NAPLs).[24] The quantitative analysis of lead alkyls is currently limited to dispensed gasolines and NAPLs, since laboratory and field experiments have demonstrated

Table 2 Selected properties of the five individual lead alkyl species, their equilibrium-predicted weight percentages in typical reacted mixtures (RMs) and physical mixtures (PMs) historically used in lead antiknock packages and selected properties of these packages (actual weight percentages varied slightly from equilibrium values and between producers).

Organic lead species	Wt.% lead	BP (°C)	TEL only	TML only	Reacted mixes			Physical mixes		
					RM25	RM50	RM75	PM25	PM50	PM75
TEL	64.06	200[a]	100.0	0.0	28.8	4.8	0.1	75.0	50.0	25.0
MTEL	66.96	178[b]	0.0	0.0	49.5	25.6	3.6	0.0	0.0	0.0
DEDML	70.14	155[b]	0.0	0.0	18.6	42.4	20.5	0.0	0.0	0.0
TMEL	73.64	133[b]	0.0	0.0	3.0	23.4	49.6	0.0	0.0	0.0
TML	77.50	110	0.0	100.0	0.1	3.8	26.2	25.0	50.0	75.0
Lead antiknock package compositions										
Wt.% Lead alkyl			61.48	50.82	58.82	56.15	53.49	58.82	56.15	53.49
Wt.% EDB			17.86	17.86	17.86	17.86	17.86	17.86	17.86	17.86
Wt.% EDC			18.81	18.81	18.81	18.81	18.81	18.81	18.81	18.81
Wt.% other[c]			1.79	12.50	4.45	7.12	9.78	4.45	7.12	9.78
Dye			0.06	0.06	0.06	0.06	0.06	0.06	0.06	0.06
Lead antiknock package properties										
Wt.% lead			39.39	39.39	39.39	39.39	39.39	39.39	39.39	39.39
Density (g ml^{-1}), 20 °C			1.586	1.583	1.587	1.585	1.854	1.583	1.583	1.583
Flash point (°C)			118	36	82	48	NA	74	54	44
VP (mmHg), 20 °C			36	40	37	39	39	38	39	40

[a]Decomposes.
[b]Interpolated.
[c]Solvent, inerts and antioxidant.

that, in the absence of free phase gasoline, lead alkyls are strongly adsorbed on soils,[44] which can inhibit their quantitative recovery.[38]

2.2.2   Oxygenates.   As governments mandated the lowering and eventual removal of lead from automotive gasoline, new classes of octane-boosting, oxygen-containing compounds, with fewer (presumed) negative impacts than lead alkyls, were being developed and used in increasing frequency to augment gasoline performance. In the 1990s, minimum oxygen contents were established for gasolines sold in certain markets with reduced air quality, thereby requiring the use of a variety of oxygenates in gasoline, *viz.* alcohols and ethers, that can provide diagnostic information on gasoline-derived contamination in the environment.

Ethanol has been the most common alcohol used in the past and its use seems destined to increase. The forensic applications of ethanol are not well established, in part from an overall lack of ethanol concentration data collected during most site assessments or other forensic studies. The analytical difficulties facing ethanol's analysis, particularly in ground water, have only recently been overcome with direct aqueous injection (DAI) techniques.[45] Thus, the database on ethanol in the environment will undoubtedly grow. However, it can be supposed that the utility of ethanol in forensic studies may be limited due to (1) the extremely rapid rate at which it is lost from gasoline upon contact with water and (2) its susceptibility to rapid biodegradation under both aerobic and anaerobic conditions.[46] Because of these properties, ethanol's presence in the environment would seem to implicate a relatively recent impact from ethanol-containing gasoline whereas its absence would be equivocal. Just how 'recent' is a function of the site-specific conditions that affect the rate of ethanol's weathering.

Numerous ethers have been used in automotive gasolines including methyl *tert*-butyl ether (MTBE), *tert*-amyl methyl ether (TAME), diisopropyl ether (DIPE) and ethyl *tert*-butyl ether (ETBE).[40] MTBE is (was) by far the most common ether – so when others are detected there is the potential for certain forensic implications. In the early 1980s, MTBE was used primarily as an octane booster within a minority of the gasoline pool, usually into the newly introduced premium unleaded blends at concentrations between 3 and 8 vol.%. By the late 1980s and throughout the 1990s, MTBE's use expanded widely worldwide with concentrations reaching up to 15 vol.%.[38] In the past few years in the USA and elsewhere, the use of MTBE has declined sharply following numerous governmental bans. The history of MTBE (or other ether oxygenate) use and concentrations in gasoline would seemingly provide a gross method of age dating gasoline-derived contamination. However, the environmental behavior of MTBE compared with hydrocarbons (*e.g.* MTBE's high aqueous solubility) and its convoluted use in some markets (*e.g.* winter only *versus* year round) can make its presence/absence or concentration difficult to interpret with respect to timing of a release. Of course, the presence of MTBE in an area that has banned MTBE indicates the presence of pre-MTBE ban gasoline. Another useful forensic application of MTBE is in terms of its concentration gradients or spatial distribution in ground water, which can reveal multiple sources/releases of the MTBE.[47] Due to the potential for co-elution with some hydrocarbons, the detection of MTBE in environmental matrices is best achieved using GC–MS in which characteristic fragment ions of MTBE (m/z 73; $C_4H_9O^+$) can be monitored.[48]

2.3 Diamondoids

Diamondoids are saturated hydrocarbons that consist of three or more fused cyclohexane rings, which results in a 'cage-like' or 'diamond-like' structure[49] – hence their class name. The simplest diamondoid, adamantane, is a tricyclo-decane with a chair configuration for each of its three cyclohexane rings. Fusion of adamantane skeletons gives rise to a series of polymantane homologues ($C_{4n+6}H_{4n+12}$, where $n \geq 1$), including diamantane, triamantane and tetra-mantane.[50] The addition of various alkyl side-chains (*e.g.* methyl and ethyl groups) yields a series of substituted diamondoids within each parent diamondoid group.

Diamondoids have been recognized in crude oils and gas condensates where they are thought to be formed during oil formation deep in the subsurface (Ref. 50 and references therein). Their specific distribution and concentration in petroleum are largely a function of the thermal history of the petroleum,[51] becoming in-creasingly enriched in gas condensates and light oils due to their inherent thermal stability.[50] Diamondoids have been demonstrated to be fairly resistant to even severe levels of biodegradation of reservoired petroleum over geological time-scales.[52,53] Their specificity and resistance to biodegradation render diamondoids potentially diagnostic compounds in fingerprinting some forms of petroleum contamination in the environment.[54,55] The distribution and concentrations of diamondoids in petroleum can be determined using GC–MS of the whole oil or the aliphatic hydrocarbon fraction of oil by monitoring well-established molecular and/or fragment ions (Table 3; Figure 2). Numerous diagnostic ratios have been developed as a means of comparing the diamondoid patterns,[56–58] and these are also useful in forensic investigations.[54] These ratios include:

MAI = 1-methyladamantane/(1-methyladamantane + 2-methyladamantane)
EAI = 2-ethyladamantane/(1-ethyladamantane + 2-ethyladamantane)
MDI = 4-methyldiamantane/(1- + 3- + 4-methyldiamantane)

In addition, Stout and Douglas[55] used diamondoids' abundance relative to C_5 alkylbenzenes as a useful metric for distinguishing highly weathered auto-motive gasoline from weathered natural gas condensate; the former of which contained much higher relative proportions of alkylbenzenes over adamantanes compared with natural gas condensate.

2.4 Acyclic Alkanes

2.4.1 n-Alkanes. n-Alkanes are often the most abundant group of hydro-carbons in crude oil and in distilled fuels. The ubiquity and susceptibility to bio-degradation of the n-alkanes can limit their utility as diagnostic compounds, but nonetheless their distributions can sometimes prove useful. Their distributions can

Table 3   Inventory of adamantane and diamantane analytes and key masses used in their identification and quantification.

Compound	Peak No.	Formula	M^+	Base peak (m/z)
Adamantane	1	$C_{10}H_{16}$	136	136
1-Methyladamantane	2	$C_{11}H_{18}$	150	135
1,3-Dimethyladamantane	3	$C_{12}H_{20}$	164	149
1,3,5-Trimethyladamantane	4	$C_{13}H_{22}$	178	163
1,3,5,7-Tetramethyladamantane	5	$C_{14}H_{24}$	192	177
2-Methyladamantane	6	$C_{11}H_{18}$	150	135
1,4-Dimethyladamantane, cis	7	$C_{12}H_{20}$	164	149
1,4-Dimethyladamantane, trans	8	$C_{12}H_{20}$	164	149
1,3,6-Trimethyladamantane	9	$C_{13}H_{22}$	178	163
1,2-Dimethyladamantane	10	$C_{12}H_{20}$	164	149
1,3,4-Trimethyladamantane, cis	11	$C_{13}H_{22}$	178	163
1,3,4-Trimethyladamantane, trans	12	$C_{13}H_{22}$	178	163
1,2,5,7-Tetramethyladamantane	13	$C_{14}H_{24}$	192	177
1-Ethyladamantane	14	$C_{12}H_{20}$	164	135
1-Ethyl-3-Methyladamantane	15	$C_{13}H_{22}$	178	149
1-Ethyl-3,5-dimethyladamantane	16	$C_{14}H_{24}$	192	163
2-Ethyladamantane	17	$C_{12}H_{20}$	164	135
Diamantane	18	$C_{14}H_{20}$	188	188
4-Methyldiamantane	19	$C_{15}H_{22}$	202	187
4,9-Dimethyldiamantane	20	$C_{16}H_{24}$	216	201
1-Methyldiamantane	21	$C_{15}H_{22}$	202	187
1,4- and 2,4-dimethyldiamantane	22	$C_{16}H_{24}$	216	201
4,8-Dimethyldiamantane	23	$C_{16}H_{24}$	216	201
Trimethyldiamantane	24	$C_{17}H_{26}$	230	215
3-Methyldiamantane	25	$C_{15}H_{22}$	202	187
3,4-Dimethyldiamantane	26	$C_{16}H_{24}$	216	201

be monitored using conventional GC–FID or by GC–MS (m/z 57, 71, or 85; $C_4H_9^+$, $C_5H_{11}^+$ or $C_6H_{13}^+$, respectively) and can reveal the boiling ranges of different types of petroleum-derived contamination.[59] In most crude oils there is a generally steady decrease in the amount of n-alkanes with increasing carbon number (*e.g.* Figure 3). However, in some crude oils in which terrigenous organic matter has contributed, there can be excess long-chain n-alkanes between C_{25} and C_{35} that exhibit a pronounced odd-carbon dominance, which has long been recognized as being characteristic of epicuticular leaf waxes.[60] The degree of enrichment of the odd-carbon n-alkanes can be measured by the carbon preference index (CPI) as[61]

$$CPI = [(C_{25} + C_{27} + C_{29} + C_{31} + C_{33})/(C_{26} + C_{28} + C_{30} + C_{32} + C_{34})$$
$$+ (C_{25} + C_{27} + C_{29} + C_{31} + C_{33})/(C_{24} + C_{26} + C_{28} + C_{30} + C_{32})]/2$$

which can be useful in some forensic investigations. However, CPIs measured in petroleum-impacted soils or sediments require caution because n-alkanes are associated with the petroleum rather than with modern, naturally occurring,

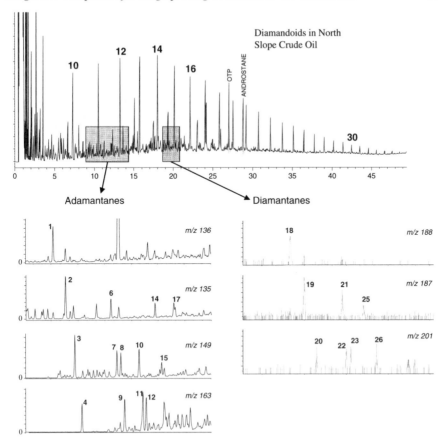

Figure 2 GC–FID of a North Slope crude oil showing the boiling ranges of adamantane and diamantanes (top) and characteristic extracted ion profiles (EIPs) of each. Compound identifications as in Table 3.

'background' plant waxes that may be present.[16] In some studies, the disparity in CPI among samples has been used to distinguish petroleum from background.[62]

The distribution of *n*-alkanes in distilled and residual fuels can aid in forensic comparisons as certain fuels are blended to contain specific distributions or proportions of *n*-alkanes. Some blended fuels can exhibit distinct and sometimes multiple maxima within the *n*-alkane 'envelope'. For example, the distribution of *n*-alkanes in diesel fuels can vary seasonally in order to maintain appropriate viscosity and the proportion of *n*-alkanes (relative to comparably boiling aromatics) in diesel fuels will affect the cetane rating.[63]

2.4.2 Isoprenoids. Isoprenoids are ubiquitous in crude oil and most refined petroleum fuels. Acyclic isoprenoids (IPs) are branched alkanes containing various combinations of isoprene units linked in either regular (head-to-tail) or irregular (tail-to-tail and head-to-head) configurations. The most commonly measured acyclic isoprenoids in oils and fuels are regular isoprenoids in the

Scott A. Stout and Zhendi Wang

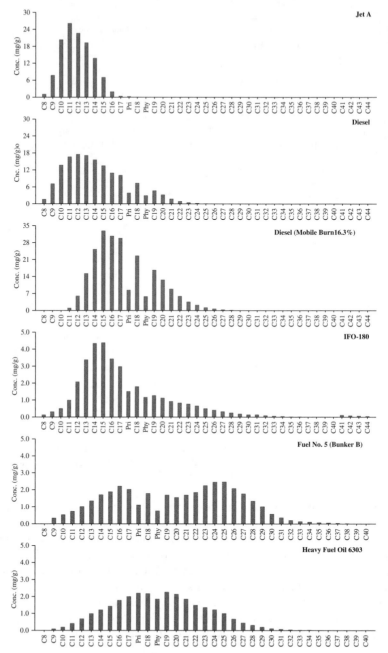

Figure 3 Quantitative distribution of *n*-alkanes in six unweathered petroleum products illustrating distinct *n*-alkane distribution patterns that reflect different blending and boiling distributions that can distinguish petroleum products in the environment. Pri, pristane; Phy, phytane.

C_{13}–C_{20} range, which includes norpristane (IP18), pristane (IP19) and phytane (IP20). While these $<$IP20 regular isoprenoids can usually be adequately measured using conventional GC–FID (provided that sufficient resolution of n-C_{17} and pristane is achieved) or GC–MS (m/z 113; $C_8H_{17}^+$), the rarer and far less abundant longer chain isoprenoids (IP21+) require GC–MS monitoring of characteristic fragment ions (*e.g.* m/z 183; $C_{13}H_{27}^+$), often in conjunction with n-alkane removal *via* fractionation.

The regular acyclic isoprenoids in the C_{13}–C_{20} range have been widely used in environmental forensic applications, particularly with middle distillate fuels that generally lack higher boiling terpane and sterane biomarkers.[64,65] Patterns or ratios among the C_{13}–C_{20} regular isoprenoids can be diagnostic, with the ratio of pristane to phytane (Pr/Ph) being perhaps the most commonly monitored 'diagnostic ratio' in petroleum and environmental geochemistry. Pr/Ph can vary among different oil sources and usually ranges from around 0.8 to 3.0,[66] although very different oil sources can share a common Pr/Ph ratio. Therefore, the Pr/Ph ratio is (alone) an inadequate metric upon which to establish a positive correlation between samples. Disparate Pr/Ph ratios, on the other hand, may prove a negative correlation when the effects of weathering (or mixing) are considered.

Chiral centers exist within most isoprenoids that gives rise to multiple diastereoisomers. Under most chromatographic conditions these diastereoisomers are not resolved, although they can be.[67] Pristane has a chiral center producing three diastereomers, *viz.* 6*R*,10*R*, 6*S*,10*S* and *meso* (or 6*R*,10*S*), that are normally present in oil at an equilibrium ratio of 1:1:2. The diagnostic utility of the isoprenoid diastereoisomers has not yet been demonstrated. However, consideration of pristane and the other isoprenoid diastereoisomers may prove diagnostic in certain forensic investigations in which other diagnostic features are absent or ambiguous. Recent work by McIntyre *et al.*[68] has described the use of the biological (*meso*) *versus* geological (6*R*,10*R* and 6*S*,10*S*) diastereoisomers of pristane as a means of assessing biodegradation of distillate fuels after the n-alkanes have been biodegraded – since the biological diastereoisomers are more susceptible.

2.5 Sesquiterpanes

Sesquiterpanes are bicycloalkanes that occur in crude oils and contain approximately 15 ('*sesqui-*') carbons. The most common bicyclic sesquiterpanes are C_{14}–C_{16} aliphatic hydrocarbons that have a decahydronaphthalene (decalin) skeleton, various C_1- and C_2-alkyl side-chains and molecular weights ranging from approximately 194 to 222 atomic mass units (amu). They were first recognized in the aliphatic fraction of a degraded Eocene crude oil from south Texas.[69] The chemical structures of the major sesquiterpanes in crude oils were subsequently shown to be based on the drimane and eudesmane skeletons.[70] The drimane-based sesquiterpanes were demonstrated to be ubiquitous in crude oils irrespective of their geological age or source rock biomass, which suggested they (like the higher boiling hopanoid biomarkers) were derived from microbial biomass.[70] Alexander *et al.*[70] also showed that eudesmane-based

Table 4 Inventory of drimane-based bicyclic sesquiterpanes showing key masses used in their identification and (relative) quantification *via* GC–MS.

Compound	Peak No.	Formula	M^+	Base peak	Confirmation ion (m/z)
Rearranged $C_{15}{}^a$	1	$C_{15}H_{28}$	208	123	193
Rearranged $C_{15}{}^a$	2	$C_{15}H_{28}$	208	123	193
8β(H)-drimane	3	$C_{15}H_{28}$	208	123	193
Rearranged $C_{15}{}^a$	4	$C_{15}H_{28}$	208	123	207
Rearranged $C_{16}{}^b$	5	$C_{16}H_{30}$	222	123	207
Rearranged $C_{16}{}^b$	6	$C_{16}H_{30}$	222	123	193
Rearranged $C_{16}{}^b$	7	$C_{16}H_{30}$	222	123	193
8β(H)-homodrimane	8	$C_{16}H_{30}$	222	123	207
Rearranged $C_{14}{}^c$	9	$C_{14}H_{26}$	194	123	179
Rearranged $C_{14}{}^c$	10	$C_{14}H_{26}$	194	123	179

[a]Unknown pentamethyldecahydronaphthalene.
[b]Unknown hexamethyldecahydronaphthalene.
[c]Unknown tetramethyldecahydronaphthalene.

sesquiterpanes were limited to oils derived from higher plant biomass and to coals. Although eudesmane-based sesquiterpanes are potentially diagnostic in some fingerprinting studies, because of their limited occurrence in terrestrial-sourced oils they are not discussed further here.

Recent studies have described the forensic application of the 10 drimane-based sesquiterpanes in environmental forensic investigations (Table 4).[71–73] These studies showed that drimane-based sesquiterpanes are ubiquitous in middle distillate fuels and that they are diagnostic, presumably inherited from the variability in the parent crude oil feedstock and/or particular refinery distillation and blending practices (Figure 4). The presence and distribution of these compounds can be observed using GC–MS upon monitoring the characteristic m/z 123 ($C_9H_{15}{}^+$) base peak and numerous confirmation ions (Table 4). The 8β(H)-drimane/[8β(H)-drimane + 8β(H)-homodrimane] ratio, the proportion of $C_{14}H_{26}$ sesquiterpanes (compounds/peaks 9 and 10, Table 4) and the relative abundances among the four $C_{15}H_{28}$ sesquiterpanes (compounds/peaks 1–4, Table 4) were diagnostic for distillate fuels from different sources and remained stable during weathering, at least through the complete removal of *n*-alkanes due to bio-degradation.[71] The crude oil literature suggests the sesquiterpanes are stable at least through acyclic isoprenoid biodegradation.[74] These combined properties, *i.e.* ubiquitous occurrence, inherent specificity and relative resistance to weathering, make the drimane-based sesquiterpanes useful as diagnostic compounds, particularly in forensic investigations involving middle distillate fuels wherein the higher boiling triterpane and sterane biomarkers are absent or nearly absent.

2.6   n-*Alkylcyclohexanes*

Cyclohexanes with a single alkyl side-chain of varying carbon numbers form a homologous series of compounds in crude oil and fuels known as

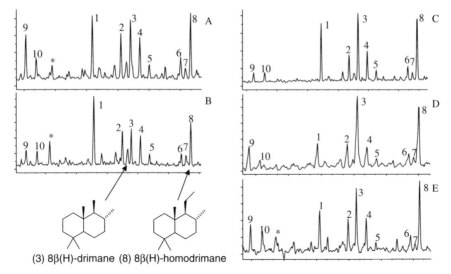

(3) 8β(H)-drimane (8) 8β(H)-homodrimane

Figure 4 Partial m/z 123 mass chromatograms for five dispensed (fresh) diesel fuel No. 2 samples demonstrating the variation in C_{14}, C_{15} and C_{16} drimane-based sesquiterpane patterns among fuels from different sources. Compound numbers refer to Table 4; chemical structures of compound Nos. 3 and 8 are shown. *, Unknown.

n-alkylcyclohexanes (CHs). The CHs range from methylcyclohexane (CH1) to around CH25. The concentrations and distributions of CHs can be determined using GC–MS upon monitoring of their characteristic fragment ion, m/z 83 ($C_6H_{11}^+$). The utility of CHs as diagnostic compounds lies in their broad boiling nature (spanning the boiling range of crude oils and petroleum products) and their greater resistance to biodegradation compared with the *n*-alkanes.[75] Like the *n*-alkanes (Section 2.4.1), the CHs can reveal the boiling range(s) of different product types and – by the presence of multiple maxima – the presence of mixtures of different petroleum products or intermediates. As an example, Figure 5 presents the GC–MS traces (at m/z 83) of Arabian Light crude oil, Jet A, Diesel Fuel No. 2 and Heavy Fuel Oil 6303, showing the characteristic distribution of *n*-alkyl-cyclohexane homologous series in these oil and oil products. For comparison, their corresponding GC–MS traces (at m/z 85) are also presented in Figure 5. CH distributions have proved useful in distinguishing the subtle boiling differences between variably blended distillate fuels containing intermediates distilled from the same crude oil feedstock (in which many other diagnostic features were comparable due to the consistent feedstock).[65] The CH alkyl side-chain has been reported to 'shorten' progressively due to a novel anaerobic biodegradation mechanism and thereby change the 'fingerprint' by lowering the boiling distribution of CHs in biodegraded fuels;[76] however this report has been questioned.[77]

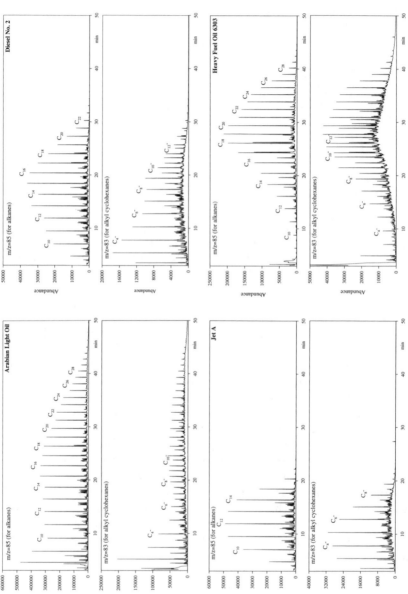

Figure 5 GC–MS extracted ion profiles of Arabian Light Oil, Jet A fuel, Diesel Fuel No. 2 and HFO 6303 illustrating distinguishing distributions of *n*-alkanes (*m/z* 85) and *n*-alkylcyclohexanes (*m/z* 83). While biodegradation may preferentially remove the *n*-alkanes, the *n*-alkylcyclohexanes can still reveal different boiling distributions of weathered petroleum in the environment.

2.7 Diterpenoids

Diterpenoids comprise a large group of natural products found throughout the plant kingdom, but particularly within higher plant resins of the conifer species. The hydrocarbons derived from these – diterpanes – are common in crude oils derived from higher plants containing resins and consist primarily of derivatives of abietane, pimarane and labdane.[78] Different crude oils exhibit distinctive distributions of tricyclic and tetracyclic diterpanes depending on their precursor plant resins.[66] Tricyclic diterpanes generally boil in the n-C_{19} to n-C_{22} range and contain approximately 20 carbons ('*di*' terpanes). They can be monitored using GC–MS using the m/z 191 ion ($C_{14}H_{23}^{+}$) and can provide diagnostic information on oils and fuels containing them.

Aromatization of the diterpenoids during diagenesis can yield numerous aromatic diterpenoids that can also provide diagnostic information in forensic investigations. Among the aromatic diterpenoids of greatest diagnostic value are retene (1-methyl-7-isopropylphenanthrene) and its probable intermediates, *e.g.* dihydroabietane, dehydroabietin and simonellite.[79] Retene can be observed using GC–MS of its molecular ion (m/z 234), which also is used to measure simultaneously the C_4-phenanthrene abundance in oils.[80] Of course, the abundance of naturally occurring 'background' retene in petroleum-impacted soils and sediments must be considered in evaluating the significance of retene in environmental samples.

2.8 Polycyclic Aromatic Hydrocarbons

Polycyclic aromatic hydrocarbons (PAHs), also called polynuclear aromatic hydrocarbons (PNAs), are nearly ubiquitous contaminants in surface soils and sediments worldwide. They are acutely toxic and have carcinogenic properties and are being recognized with increasing frequency as major contaminants in sediments, particularly in urban environments. Determining the source(s) of PAHs in soils and sediments is a frequent forensic task that requires a basic understanding of their chemistry and nomenclature and formation.

The PAHs, as their name implies, (1) contain multiple 'ring' structures, which are (2) aromatic in nature and (3) comprised of hydrogen and carbon. The arrangement and number of rings are used to distinguish different PAHs. Historically, the PAHs of principal environmental concern, as designated by the US EPA, are those 16 listed in the EPA's Priority Pollutant List. However, hundreds of PAHs exist and many of these have proven essential to forensic investigations addressing PAH sources (Ref. 81 and references therein). For example, all of the priority pollutant PAHs are 'non-alkylated', *i.e.* they contain no carbon side-chains. Non-alkylated PAHs are sometimes called 'parent' or C_0-PAHs. However, many 'alkylated' PAHs exist that contain carbon side-chains of varying number, length and location. These C_1–C_4-PAHs are not on the EPA's Priority Pollutant List and therefore are not included in the standard EPA Method 8270. However, it is both the parent and alkylated PAHs (C_0–C_4-PAHs; Table 5) that when combined are most useful in environmental forensic investigations.

Table 5 Inventory of parent and alkylated PAHs and heterocyclic aromatic compounds often used in chemical fingerprinting studies.[a]

Low molecular weight (LPAH)

Target PAH/PAH group	Key	Quantification ion (m/z)	RF
Naphthalene	N0	128	N0
C1-Naphthalenes	N1	142	N0
C2-Naphthalenes	N2	156	N0
C3-Naphthalenes	N3	170	N0
C4-Naphthalenes	N4	184	N0
Acenaphthene	ACE	154	ACE
Acenaphthylene	ACY	152	ACY
Biphenyl	BPHN	154	BPHN
Dibenzofuran	DBF	168	DBF
Fluorene	F0	166	F0
C1-Fluorenes	F1	180	F0
C2-Fluorenes	F2	194	F0
C3-Fluorenes	F3	208	F0
Dibenzothiophene	D0	184	D0
C1-Dibenzothiophenes	D1	198	D
C2-Dibenzothiophenes	D2	212	D
C3-Dibenzothiophenes	D3	226	D
C4-Dibenzothiophenes	D4	240	D
Anthracene	AN	178	AN
Phenanthrene	P0	178	P0
C1-Phenanthrenes/anthracenes	P1	192	P0
C2-Phenanthrenes/anthracenes	P2	206	P0
C3-Phenanthrenes/anthracenes	P3	220	P0
C4-Phenanthrenes/anthracenes	P4	234	P0

High molecular weight (HPAH)

Target PAH/PAH groups	Key	Quantification ion (m/z)	RF
Fluoranthene	FL	202	FL
Pyrene	PY	202	PY
C1-Fluoranthenes/pyrenes	FP1	216	FL
C2-Fluoranthenes/pyrenes	FP2	230	FL
C3-Fluoranthenes/pyrenes	FP3	244	FL
C4-Fluoranthenes/pyrenes	FP4	258	FL
Naphthobenzothiophenes	NT0	234	NT0
C1-Naphthobenzothiophenes	NT1	248	NT0
C2-Naphthobenzothiophenes	NT2	262	NT0
C3-Naphthobenzothiophenes	NT3	276	NT0
C4-Naphthobenzothiophenes	NT4	290	NT0
Benzo[a]anthracene	BaA	228	BaA
Chrysene	C0	228	C0
C1-Chrysenes	C1	242	C0
C2-Chrysenes	C2	256	C0
C3-Chrysenes	C3	270	C0
C4-Chrysenes	C4	284	C0
Benzo[b]fluoranthene	BbF	252	BBF
Benzo[j,k]fluoranthene	BjkF	252	BKJF
Benzo[a]fluoranthene	BaF	252	BAF
Benzo[a]pyrene	BaP	252	BAP
Benzo[e]pyrene	BeP	252	BEP
Perylene	PER	252	PER
Indeno[1,2,3-cd]pyrene	IND	276	IND
Dibenzo[a,h]anthracene	DA	278	DA
Benzo[ghi]perylene	GHI	276	GHI

[a]Compounds in bold are US EPA Priority Pollutant PAHs. The quantification ions and the response factor (RF) compounds are given for each analyte.

The source(s) of PAHs in the environment was first investigated in the early-to mid-1970s, when their presence was of growing interest and concern. Early investigators were confounded regarding the origin of PAHs, due in part to analytical constraints. The breakthrough in determining the origins of PAHs in sediments came when researchers investigated both non-alkylated and alkylated PAH distributions,[82,83] rather than individual, non-alkylated PAHs. Researchers quickly determined that the PAHs in the environment were derived predominantly from two different anthropogenic sources (*i.e.* combusted/pyrolyzed fossil fuel *versus* spilled petroleum, or 'pyrogenic' *versus* 'petrogenic'). These two source categories could be readily distinguished on the basis of their alkyl PAH distributions.[82–84] This is discussed in Section 2.8.1. Non-anthropogenic sources of PAHs (*e.g.* naturally occurring PAHs) derived from natural fires or biogenic sources were also recognized to occur in some urban and coastal sediments.[82–85]

Measuring the PAH distribution in petroleum and petroleum-impacted matrices is performed using a modification of the EPA Method 8270, *Semi-volatile Organic Compounds by Gas Chromatography/Mass Spectrometry (GC–MS)*, as detailed by NOAA Status and Trends methodology[86] and in the Federal Register (Federal Register 40CFR Subchapter J, Part 300, Subpart L, Appendix C100, par. 4.6.3 to 4.6.5).

In this method, the gas chromatograph is operated with a very slow oven temperature program to optimize the separation of target compounds (Table 5) and the mass spectrometer is operated in the selected ion monitoring (SIM) mode to minimize interferences from non-target compounds and, when necessary, to improve the detection limits of low-concentration analytes.[24,87] Internal surrogate and recovery standards are used to measure performance and concentrations, relative to an external calibration solution containing the parent PAHs on the analyte list. The response factors (RFs) for the parent PAHs are applied to the appropriate alkylated PAHs (Table 5).

2.8.1   Petrogenic versus Pyrogenic.   Because of the nature of their formation and similar physical/chemical properties, groups of petrogenic or pyrogenic PAHs tend to co-occur in the environment. This knowledge allows the investigator to recognize specific PAH assemblages, or 'fingerprints', as being derived from a certain source. PAH source recognition studies require the identification of both non-alkylated and alkylated PAHs, in addition to selected heterocyclics (Ref. 81 and references therein), in order to distinguish various PAH sources, including urban background.[88]

Pyrogenic and petrogenic PAHs can be readily distinguished on the basis of their alkyl group distributions.[85] Figure 6 shows the basic relative distributions of variously alkylated PAHs in distinct and variably weathered sources. The petrogenic PAH homologue profiles present in crude oil and diesel fuel form a characteristic 'bell-shaped' pattern due to the relative abundance of alkylated PAHs (C_2- and C_3-PAHs) over the corresponding parent PAHs. Pyrogenic PAH homologue profiles present in coal tar and creosote form a decreasing or

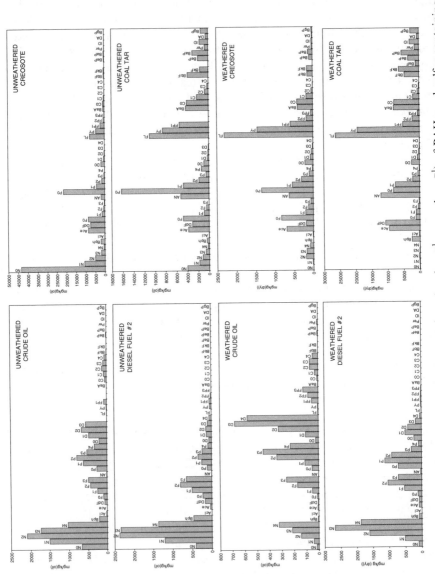

Figure 6 Histograms showing the distributions and concentrations (mg kg$_{oil}^{-1}$ or mg kg$_{dry}^{-1}$) of PAHs and sulfur-containing aromatics in fresh and weathered crude oil, Diesel Fuel No. 2, coal tar and creosote demonstrating the varying profiles for petrogenic and pyrogenic PAH sources. Compound identification as in Table 5.

'sloped' pattern due to the domination of the parent PAH (C_0) and decreasing abundance of each increasingly alkylated PAH homologue. Weathering tends to affect the profiles for the low molecular weight PAHs (LPAHs) (naphthalenes and phenanthrenes), in which the susceptibility decreases with increasing degree of alkylation, whereas those of the high molecular weight PAHs (HPAHs) remain mostly stable.[5,89]

Numerous diagnostic ratios have been proposed to differentiate pyrogenic PAHs from other hydrocarbon sources, each exploiting the aforementioned disparity in the degree of alkylation.[90–95] Wang *et al.*[95] proposed a 'Pyrogenic Index' (PI) as a quantitative indicator for the characterization of pyrogenic PAHs and for the differentiation of pyrogenic and petrogenic PAHs. The PI was defined as the ratio of the total concentration of EPA Priority Pollutant PAHs containing 3–6 rings (Table 5) to the total concentration of five alkylated PAH homologues:

$$PI = \Sigma(3 - 6\text{-ring EPA PAHs})/\Sigma(5 \text{ alkylated PAHs})$$

Compared with other diagnostic ratios obtained from individual compounds, the advantages of this index include: (1) petrogenic and pyrogenic PAHs are characterized by the dominance of five alkylated PAH homologous series and by the dominance of unsubstituted high molecular weight PAHs, respectively; therefore, changes in this ratio more truly reflect the difference in the PAH distribution between these two sets of hydrocarbons; (2) better accuracy with less uncertainty than ratios reliant on individual PAH compounds; (3) high degree of consistency from sample to sample and little interference from the concentration fluctuation of individual components within the PAH series; and (4) the long-term natural weathering and biodegradation only slightly alter the values of this ratio, but the ratio will be dramatically altered by combustion. Therefore, the PI can be used as a general and effective criterion to differentiate unambiguously pyrogenic and petrogenic PAHs.[95]

The PI *versus* Ph/An values for over 100 oils and refined products have been reported.[95] It was found that lighter refined products and most crude oils show PI ratios less than 0.01 whereas heavy oils and heavy fuels show higher PI ratios, *viz.* 0.01–0.05. The PI increases dramatically up to much higher values for pyrogenic materials (for example, it increased to 0.8–2.0 for the 1994 Mobile burn soot samples). The usefulness of the PI in environmental forensic investigations for input of pyrogenic PAHs and spill source identification has been clearly demonstrated in several recent spill case studies[96–98] and a tyre fire case study.[99]

2.8.2 Petrogenic Specific Isomers. The relative abundances of various individual PAH isomers have been used in studies addressing the sources of PAHs in the environment.[100–105] The basis for this approach lies with the relative kinetic stabilities of different PAH isomers during their formation under different heating/cooling conditions and by their comparable environmental fate.[103,106] Of specific interest to petroleum are the studies of Radke and coworkers that have demonstrated the relative stability of alkylated naphthalene and phenanthrene isomers depend upon the nature of the geological heating experienced by petroleum source rocks and crude oils.[107] Various ratios have demonstrated the relative instability of methyl- and dimethylnaphthalene (MN and DMN) and methylphenanthrene (MP) isomers with alkyl groups in the α-position *versus* those with alkyl groups in the β-position, such that with increased thermal stress there is a shift from α-type to β-type alkylation.[108,109] These ratio include:

MNR = 2-MN/1-MN
DNR = (2,6-DMN + 2,7-DMN)/1,5-DMN
MPI 1 = 1.5(2-MP + 3-MP)/(P + 1-MP + 9-MP)
MPI 2 = 3(2-MP)/(P + 1-MP + 9-MP)
MPR = 2-MP/1-MP

The proportions or concentrations of individual methyl-naphthalene (m/z 142), C_2-naphthalene (m/z 156), phenanthrene (m/z 178), methylphenanthrene (m/z 192) and methyldibenzothiophene (m/z 198) isomers can used to calculate the Radke-type 'maturity' ratios (above) that can prove diagnostic between crude and refined petroleum of different origins. Thus, whereas the overall degree of alkylation is more important in distinguishing petrogenic from pyrogenic PAHs (Figure 6), the individual PAH isomer distributions can be more important in distinguishing one source of petroleum from another. Some caution is necessary when applying conventional PAH-based isomer ratios to environmental samples in which preferential biodegradation of particular isomer(s) over others can affect such ratios, independent of the petroleum source. For example, the susceptibility among the methylphenanthrene (MP) isomers was demonstrated to be 3-MP > 2-MP > 1-MP > 9-MP in crude oil during degradation.[110]

2.8.3 Sulfur-containing Aromatics. Sulfur is usually the most abundant heterocyclic compound found in petroleum. Sulfur-containing aromatic compounds are a special class of heterocyclics that include the thiophenes, benzothiophenes, dibenzothiophenes, naphthothiophenes, naphthobenzothiophenes and several larger compounds. The analysis of sulfur-containing aromatics in crude oil and petroleum products can be achieved using numerous selective detectors (*e.g.* flame photometric, atomic emission, sulfur chemiluminescence and mass spectrometers). The abundance and distribution of sulfur-containing aromatics in crude oils is a function of the geological history, which at higher levels can prove problematic during the refining process (catalyst poisoning) and in terms of emissions from the refined fuels. Although automotive gasoline can contain numerous thiols, sulfides and thiophenes that have some diagnostic value,[111,112] most chemical fingerprinting studies that have relied upon the relative abundance of sulfur-containing aromatics to distinguish different petroleum sources have focused on distillate and residual fuels and crude oils. For example, Wang and Fingas[113] showed that the proportions of methyldibenzothiophene (MDBT) isomers (1-MDBT and 2-/3-MDBT relative to 4-MDBT) varied among different crude oil and residual fuel sources. However, preferential biodegradation of the

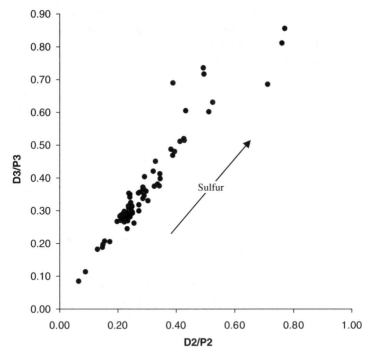

Figure 7 Double ratio plot of D2/P2 and D3/P3 in residual fuel oil samples (IFO380) (after Uhler *et al.*[119]).

2-/3-MDBT can occur rendering the 1-MDBT/4-MDBT ratio most useful for source differentiation.[110] The abundances of C_2- and C_3-alkylated dibenzothiophenes relative to C_2- and C_3-alkylated phenanthrenes (C_2-DBT/C_2-P and C_3-DBT/C_3-P) has been shown to vary among distillate and residual fuels and crude oils,[114] and to remain constant during weathering.[115] As such, double ratio plots of D2/P2 *versus* D3/P3 have been used to describe the varying abundance of sulfur-containing aromatics in fuels and environmental samples (Figure 7).

In addition to reducing the overall abundance of sulfur in fuels, the refining process can have very specific effects on certain sulfur-containing aromatic isomers over others.[116] Specifically, DBTs with methyl or ethyl groups in the 4-position are highly resistant to hydrodesulfurization (HDS) and tend to predominate in HDS-processed fuels. This type of refining-induced variation has not yet been applied to environmental forensic investigations, although its potential has been recognized.[117]

2.9 Triterpenoids

Terpenoids are a class of hydrocarbons which, like acyclic isoprenoids (Section 3.4.2), are comprised of isoprene [2-methyl-1,3-butadiene, $CH_2 = C(CH_3)$ $CH = CH_2$] subunits. Terpenoids can be classified according to the number of isoprene units from which they are biogenetically derived, even though some carbons may have been added or lost.[118] The saturated terpenoids that occur in petroleum are generally categorized into families based on the approximate number of isoprene subunits they contain. Terpenoids containing 2–8 isoprene subunits are termed mono- (C_{10}), sesqui- (C_{15}), di- (C_{20}), sester- (C_{25}), tri- (C_{30}) and tetra- (C_{40}) terpenoids. The saturated terpenoids – called terpanes – comprise several homologous series, including bicyclic, tricyclic, tetracyclic and pentacyclic compounds that are found in crude oil. The diagnostic value of the sesquiterpanes and diterpanes in petroleum fingerprinting studies was discussed previously (Sections 2.5 and 2.7, respectively). Although sester-, tri- and tetra- terpenoids containing almost any number of carbons can occur in theory, the triterpenoids containing combinations of five- or six-membered carbon rings (cyclopentyl or cyclohexyl) occur most commonly in petroleum and are the focus of this section. Figure 8 shows the molecular structures of some repre- sentative triterpanes discussed here.

Hopanes are pentacyclic triterpanes commonly containing 27–35 carbon atoms in a naphthenic structure composed of four six-membered rings and one five-membered ring (Figure 8). The origins of most hopanes are the C_{35} bac- teriohopanetetrol (tetrahydroxybacteriohopane) and related bacteriohopanes found in the lipid membranes of prokaryotic organisms.[120] Therefore, hopanes are ubiquitous in sediments, source rocks and ultimately crude oil. Hopanes with the 17α(H),21β(H)-configuration in the range C_{27}–C_{35} are often charac- teristic of petroleum because of their high abundance and thermodynamic stability compared with other epimeric (ββ and βα) series. Table 6 lists the important terpanoids frequently used in the chemical fingerprinting of petroleum.[4]

Tricyclic   **Tetracyclic**

C_{30} tricyclic terpane C_{24} 17,21-secohopane
(R=$C_{12}H_{25}$, an extended cheillanthane)

Pentacyclic

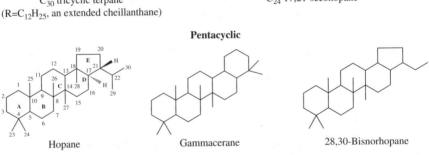

Hopane   Gammacerane   28,30-Bisnorhopane

Figure 8 Molecular structures of selected tricyclic, tetracyclic and pentacyclic tri- terpanes whose abundance can be diagnostic for different petroleum sources.

Table 6 Tri-, tetra- and pentacyclic triterpane biomarkers frequently used for chemical fingerprinting of petroleum in the environment.

Peak No.	Compound	Key	Empirical formula	Target ion (m/z)
1	C_{19} tricyclic terpane	TR19	$C_{19}H_{34}$	191
2	C_{20} tricyclic terpane	TR20	$C_{20}H_{36}$	191
3	C_{21} tricyclic terpane	TR21	$C_{21}H_{38}$	191
4	C_{22} tricyclic terpane	TR22	$C_{22}H_{40}$	191
5	C_{23} tricyclic terpane	TR23	$C_{23}H_{42}$	191
6	C_{24} tricyclic terpane	TR24	$C_{24}H_{44}$	191
7	C_{25} tricyclic terpane (a)	TR25A	$C_{25}H_{46}$	191
8	C_{25} tricyclic terpane (b)	TR25B	$C_{25}H_{46}$	191
9	Triplet: C_{24} tetracyclic terpane + C_{26} ($S + R$) tricyclic terpanes	TET24 + TR26A + TR26B	$C_{24}H_{42}$ + $C_{26}H_{48}$	191
10	C_{28} tricyclic terpane (a)	TR28A	$C_{28}H_{52}$	191
11	C_{28} tricyclic terpane (b)	TR28B	$C_{28}H_{52}$	191
12	C_{29} tricyclic terpane (a)	TR29A	$C_{29}H_{54}$	191
13	C_{29} tricyclic terpane (b)	TR29B	$C_{29}H_{54}$	191
14	Ts: 18α(H),21β(H)-22,29,30-trisnorhopane	Ts	$C_{27}H_{46}$	191
15	17α(H),18α(H),21β(H)-25,28,30-Trisnorhopane	TH27	$C_{27}H_{46}$	191, 177
16	Tm: 17α(H),21β(H)-22,29,30-trisnorhopane	Tm	$C_{27}H_{46}$	191
17	C_{30} tricyclic terpane 1	TR30A	$C_{30}H_{56}$	191
18	C_{30} tricyclic terpane 2	TR30B	$C_{30}H_{56}$	191
19	17α(H),18α(H),21β(H)-28,30-Bisnorhopane	H28	$C_{28}H_{48}$	191, 163
20	17α(H),21β(H)-25-Norhopane	NOR25H	$C_{29}H_{50}$	191, 177
21	17α(H),21β(H)-30-Norhopane	H29	$C_{29}H_{50}$	191
22	18α(H),21β(H)-30-Norneohopane (C_{29}Ts)	C29Ts	$C_{29}H_{50}$	191
23	17α(H)-Diahopane	DH30	$C_{30}H_{52}$	191
24	17α(H),21β(H)-30-Norhopane (normoretane)	M29	$C_{29}H_{50}$	191
25	18α(H) and 18β(H)-oleanane	OL	$C_{30}H_{52}$	191, 412
26	17α(H),21β(H)-Hopane	H30	$C_{30}H_{52}$	191
27	17α(H)-30-Nor-29-homohopane	NOR30H	$C_{30}H_{52}$	191
28	17β(H),21α(H)-Hopane (moretane)	M30	$C_{30}H_{52}$	191
29	22S-17α(H),21β(H)-30-Homohopane	H31S	$C_{31}H_{54}$	191
30	22R-17α(H),21β(H)-30-Homohopane	H31R	$C_{31}H_{54}$	191
31	Gammacerane	GAM	$C_{30}H_{52}$	191, 412
32	17β(H),21β(H)-Hopane	(IS)	(Internal standard)	191
33	22S-17α(H),21β(H)-30,31-Bishomohopane	H32S	$C_{32}H_{56}$	191
34	22R-17α(H),21β(H)-30,31-Bishomohopane	H32R	$C_{32}H_{56}$	191
35	22S-17α(H),21β(H)-30,31,32-Trishomohopane	H33S	$C_{33}H_{58}$	191

Table 6   (*Continued*).

Peak No.	Compound	Key	Empirical formula	Target ion (m/z)
36	22R-17α(H),21β(H)-30,31,32-Trishomohopane	H33R	$C_{33}H_{58}$	191
37	22S-17α(H),21β(H)-30,31,32,33-Tetrakishomohopane	H314S	$C_{34}H_{60}$	191
38	22R-17α(H),21β(H)-30,31,32,33-Tetrakishomohopane	H34R	$C_{34}H_{60}$	191
39	22S-17α(H),21β(H)-30,31,32,33,34-Pentakishomohopane	H35S	$C_{35}H_{62}$	191
40	22R-17α(H),21β(H)-30,31,32,33,34-Pentakishomohopane	H35R	$C_{35}H_{62}$	191

These compounds can be detected in low quantities (ppm and sub-ppm levels) in petroleum and petroleum-impacted matrices by the use of the GC–MS operated in the SIM mode. For better resolution and improved sensitivity, the silica gel column clean-up and fractionation technique is often applied to oil samples prior to the GC–MS measurement. Saturated biomarkers are eluted and collected in the saturated fraction with other saturated hydrocarbons, whereas aromatic steranes are eluted in the aromatic fraction with other aromatic compounds, including alkylated PAHs.[24,32] Two types of calibration standards are appropriate for the determination of biomarker concentration. The first are retention time marker compounds [*e.g.* 5β(H)-cholane], against which measured biomarker retention times can be compared.[121] The second are discrete biomarker chemicals, available commercially, that can be used to generate relative response factors (RRFs) for individual compounds and compounds of similar structure.

The chemical structures of terpenoids are more complicated than those of *n*-alkanes and acyclic isoprenoids, particularly due the three-dimensional nature of the terpenoid structure. The system used for the nomenclature of terpenoids has evolved over many years. For many terpenoid classes, several names have been proposed for the carbon skeleton, but the basic rules of the IUPAC system are used for the nomenclature of biomarkers. Cyclic triterpanes are labeled according to the following rules.[122,123] (1) Each carbon atom and the rings in biomarker molecules are labeled systematically. Rings are specified in succession from left to right as the A-ring, B-ring, C-ring, D-ring and so on. (2) A capital C followed immediately by a subscript number refers to the number of carbon atoms in a particular compound (*e.g.* C_{30} hopane and C_{27} sterane mean that they contain 30 and 27 carbon atoms, respectively). (3) A capital C followed by a dash and numbers refers to a particular position within the compound [*e.g.* C-17 and C-21 in the 17α(H),21β(H)-hopane are the carbon atoms at positions 17 and 21]. (4) Prefixes are used to indicate the changes to the normal biomarker carbon skeleton, which include the prefixes *nor-, seco-, neo-* and others.

Table 7 summarizes the nomenclature used to modify the structural specification of all cyclic biomarkers, including the terpenoids. The prefix *nor-* is used to indicate loss of carbons from a carbon skeleton. For example, $17\alpha(H),21\beta(H)$-30-norhopane is identical with C_{30} $17\alpha(H),21\beta(H)$-hopane (Figure 8) except that a methyl group at the C-30 position has been lost from its

Table 7 Common modifiers and nomenclature used to modify the structural specification of cyclic biomarkers (modified from Peters and Moldowan[122]).

Modifier	Description	Example biomarker
Homo-	One additional carbon on the parent molecular structure	C_{31} $17\alpha(H),21\beta(H)$-30-homohopane
Bis-, tris-, tetrakis-, pentakis- (also di-, tri-, tetra-, penta-)	Two to five additional carbons on the parent molecular structure	C_{32} $17\alpha,21\beta$-30,31-bishomohopane
		C_{33} $17\alpha,21\beta$-30,31,32-trishomohopane
		C_{34} $17\alpha,21\beta$-30,31,32,33-tetrakishomohopane
		C_{35} $17\alpha,21\beta$-30,31,32,33,34-pentakishomohopane
Seco-	Cleaved C–C bond	C_{24} 17,21-secohopane (tetracyclic)
Nor-	One less carbon on the parent molecular structure	25-Norhopane
Bisnor-	Two less carbons on the parent molecular structure	28,30-Bisnorhopane
Trisnor-	Three less carbons on the parent molecular structure	25,28,30-Trisnorhopane
Neo-	Methyl group shifted from C–18 to C–17 position on hopanes	C_{29}Ts: 30-norneohopane
α	Asymmetric carbon in ring with 'H' down	$17\alpha(H),21\beta(H)$-Hopane
β	Asymmetric carbon in ring with 'H' up	$17\beta(H),21\beta(H)$-Hopane
R	Asymmetric carbon in acyclic moiety of biomarkers obeying convention in a clockwise direction	C_{27} 20*R*-cholestane
S	Asymmetric carbon in acyclic moiety of biomarkers obeying convention in a clockwise direction	C_{27} 20*S*-cholestane

point of attachment at the C-22 position. Similarly, 25-norhopanes are identical with C_{30} hopane except that a methyl group (C-25) has been removed from its point of attachment at the C-10 position. If two or three carbons are lost, the prefix *bisnor-* or *trisnor-* is used, respectively. Thus, 28,30-bisnorhopanes have two methyl groups (C-28 and C-30) removed from their parent C_{30} hopane (Figure 8). The prefix *seco-* is used to indicate cleavage of a bond, with the locants for both ends of the broken bond given, *e.g.* 3,4-secoeudesmane. The 17,21-secohopane indicates that the bond between carbon numbers 17 and 21 in the E-ring of C_{30} hopane has been broken, resulting in the formation of a new tetracyclic terpane (Figure 8). The prefix *homo* is used to indicate addition of a carbon from the parent carbon skeleton, for example, 30-homohopanes are identical with C_{30} hopane except that a methyl group has been added at the C-30 position. If two to five carbons are added on the parent molecular structure, the prefixes *bis, tris, tetrakis* and *pentakis* are used, respectively. A brief description of α and β and *R* and *S* stereoisomers is also included in Table 7.

Triterpanes clearly meet the criteria of diagnostic compounds (Section 2) given their relative resistance to weathering and specificity. Therefore, chemical fingerprinting of petroleum has utilized numerous diagnostic ratios involving the triterpanes (Table 8) for determining the source of spilled oil, differentiating and correlating oils, monitoring the degradation process and weathering state of oils under a wide variety of conditions[42,124–134] and in indicating chronic industrial and urban releases.[42,135,136] It is important to realize that the suit of diagnostic ratios listed in Table 8 [which can be either calculated from quantitative (*i.e.* compound concentrations) or semiquantitative data (*i.e.* peak areas or heights)] is neither inclusive nor appropriate for all oil spill identification cases. In some spill cases, it may be prudent to include some particularly characteristic ratios. In other situations, the abundance of some biomarkers may be too low to obtain reliable diagnostic ratios. Thus, maintaining flexibility in the selection of diagnostic ratios to be used in specific cases is important.

2.9.1 Extended Tricyclic and Tetracyclic Triterpanes.   Terpanes are found in nearly all oils and oils from different source rocks deposited under different conditions may show different terpane fingerprints. As an example, Figure 9 shows the SIM chromatograms for common biomarker classes in a Kuwait crude oil obtained by using a 60 m column.[137] Thirty-eight terpanes from C_{19} tricyclic terpane to C_{35} homohopanes (m/z 191) and 19 steranes from C_{21}–C_{29} steranes (m/z 217 and 218) in total have been unambiguously identified and characterized in this oil.

A series of tricyclic terpanes ranging from C_{19} to C_{30} are found in most oils and bitumens[138,139] and they are highly resistant to biodegradation. It can be seen from Figure 9 that the tricyclic terpanes from C_{19} to C_{30} plus the C_{24} tetracyclic terpane are well resolved and clearly recognized in this Kuwait oil. Tricyclic terpanes are unrelated to pentacyclic hopanes because of their

Table 8 Diagnostic ratios among triterpanes that are frequently used for identification, correlation and differentiation of spilled oils (after Wang *et al.*[32]).

Diagnostic ratio	Code
C_{21}/C_{23} tricyclic terpane	TR21/TR23
C_{23}/C_{24} tricyclic terpane	TR23/TR24
C_{23} tricyclic terpane/C_{30} $\alpha\beta$-hopane	TR23/H30
C_{24} tricyclic terpane/C_{30} $\alpha\beta$-hopane	TR24/H30
[C_{26} (*S*) + C_{26} (*R*) tricyclic terpane]/C_{24} tertracyclic terpane	Triplet ratio
C_{27} 18α,21β-trisnorhopane/C_{27} 17α,21β-trisnorhopane	Ts/Tm
C_{28} bisnorhopane/C_{30} $\alpha\beta$-hopane	H28/H30
C_{29} $\alpha\beta$-25-norhopane/C_{30} $\alpha\beta$-hopane	NOR25H/H30
C_{29} $\alpha\beta$-30-norhopane/C_{30} $\alpha\beta$-hopane	H29/H30
Oleanane/C_{30} $\alpha\beta$-hopane	OL/H30
Moretane(C_{30} $\beta\alpha$ hopane)/C_{30} $\alpha\beta$-hopane	M30/H30
Gammacerane/C_{30} $\alpha\beta$-hopane	GAM/H30
Tricyclic terpanes (C_{19}-C_{26})/C_{30} $\alpha\beta$-hopane	Σ(TR19–TR26)/H30
C_{31} homohopane (22*S*)/C_{31} homohopane (22*R*)	H31S/H31R
C_{32} bishomohopane (22*S*)/C_{32} bishomohopane (22*R*)	H32S/H32R
C_{33} trishomohopane (22*S*)/C_{33} trishomohopane (22*R*)	H33S/H33R
Relative homohopane distribution	H31:H32:H33:H34:H35
Σ(C_{31}–C_{35})/C_{30} $\alpha\beta$-hopane	Σ(H31–H35)/H30
Homohopane index	H31/Σ(H31–H35) to H35/Σ(H31–H35)

different point of attachment of side-chains (see Figure 8). Based on structure studies,[140] the C_{24}–C_{27} tetracyclic terpanes appears to be degraded hopanes (such as 17,21-secohopanes; Figure 8), and these tetracyclic terpanes appear to be more resistant to biodegradation and maturation than the hopanes. The C_{24} tetracyclic terpane shows the most widespread occurrence, followed by C_{25}–C_{27} homologues.

Palacas *et al.*[141] found that tricyclic terpanes were the most useful series of biomarkers for differentiating potential from effective source rocks in the South Florida Basin. The diagnostic ratios of paired tricyclic terpanes (*e.g.* C_{23}/C_{24} and triplet ratios) and tricyclics/C_{30}-hopanes have been widely used for forensic oil spill investigation.[32] The triplet ratio (Table 8) was first used by Kvenvolden *et al.*[142] to study a North Slope crude, in which the ratio is ~2. The spilled Exxon Valdez oil (an Alaska North Slope crude) and its residues also have triplet ratios of ~2.[126] Conversely, many tarballs and residues collected from the shorelines of the Prince William Sound were similar to each other but chemically distinct from the spilled Exxon Valdez oil with triplet ratios of ~5. The triplet ratio, combined with other diagnostic biomarker ratios and isotopic compositions, revealed that these non-Valdez tarballs originated from California with a likely source being the Monterey Formation.[126] During the

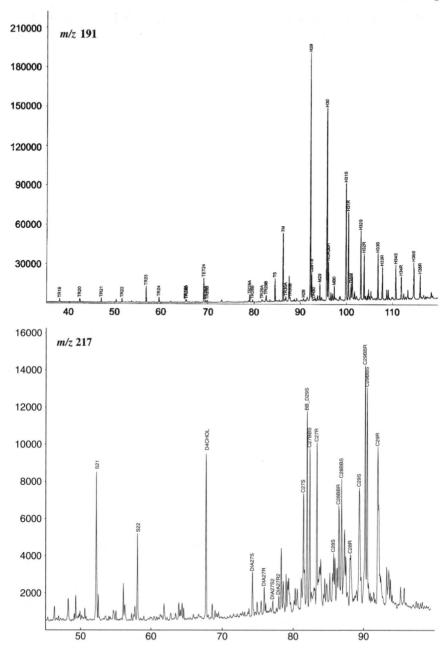

Figure 9 Partial GC–MS EIPs showing m/z 191 and 217 of a Kuwait crude oil illustrating the common terpane and sterane biomarker classes, respectively. See Tables 6 and 9 for compound identifications.

Arrow oil spill work, the ratio of the most abundant C_{29}-C_{30} hopane and also C_{23}/C_{24} and other biomarker ratios including C_{23}/C_{30} and C_{24}/C_{30} hopane were defined and used by Wang *et al.*[132] as reliable source indicators.

2.9.2 Pentacyclic Triterpanes – Hopanoids. Among the most common and useful terpanes found in petroleum is a homologous suite of pentacyclic triterpanes (C_{27}-C_{35}) known as hopanes (Table 6) that is evident in the m/z 191 mass chromatogram due to a strong fragment ion derived from rings (A + B) of the hopane molecule, but rings (D + E) may also be the source. The m/z 177 fragment is most likely derived from rings (A + B) of hopane molecules which have lost a methyl group from position 10, that is, 25-norhopanes.[143–145] The notable feature of the mass spectra of 25-demethylated hopanes is that the m/z 177 fragment has higher intensity than the m/z 191 fragment. (The regular hopanes also give the m/z 177 fragment upon electron ionization in the mass spectrometer, but in much lower abundance than the m/z 191 fragment.) The m/z 177 fragment is formed by the loss of CH_2 from the m/z 191 fragment and can be seen in all mass spectra of hopanes. The biomarker C_{29} 18α(H),21β(H)-30-norneohopane (or C_{29}Ts), which elutes immediately after C_{29} $\alpha\beta$-hopane, has a greater abundance of the m/z 177 ion than of the m/z 191 ion. C_{29} 17β(H),21α(H)-30-norhopane (normoretane) also shows significantly greater abundance of the m/z 177 ion than of the m/z 191 ion.

Depending on the oil sources, depositional environment and geological migration conditions, oils can have large differences in distribution patterns and concentrations of triterpanes.[4,32] In most cases, characterization of triterpane biomarkers should include the determination of both absolute concentrations and relative fingerprinting distributions and should not be just measuring peak ratio alone. This is important because it is possible to have a situation where a source might have a similar biomarker ratio but very different actual amounts of biomarkers. Quantitative determination of biomarkers is also critical in oil spill studies involving recognition and/or allocation of mixtures of different oils.[30] For a given type of organic material, the biomarker concentrations generally decrease with increasing thermal maturity. Very light oils or condensates (*e.g.* the Scotia Light) typically contain low concentrations of detectable hopanes. The Alaska North Slope (ANS) oil contains a wide range of terpanes from C_{20} tricyclic terpane to C_{35} pentacyclic terpanes with the C_{30} $\alpha\beta$-hopane as the most abundant, followed by C_{29} $\alpha\beta$-hopane. The triplet C_{24} tetracyclic + C_{26} (S + R) tricyclic terpanes are also highly abundant. In contrast, the Arabian Light, South Louisiana and Troll oils have terpanes largely located in the C_{27} to C_{35} pentacyclic hopane range and contain only small amounts of C_{20}-C_{24} tricyclic terpanes. In addition, the abundance of C_{29} $\alpha\beta$-hopane is higher than that of C_{30} $\alpha\beta$-hopane in Arabian Light crude oil. The dominance of C_{28} 17α(H),18α(H),21β(H)-28,30-bisnorhopane (BHN28) is particularly prominent in California API-11, Sockeye and Platform Elly (all three oils are from California), and its abundance is even higher than C_{30} and/or C_{29} 17α(H),21β(H)-hopane. A high concentration of C_{28} 17α(H), 18α(H),21β(H)-28,30-bisnorhopane is typical of petroleum from highly reducing to anoxic depositional environments.[146] The California API-11 and

Platform Elly demonstrate higher concentrations of C_{31}–C_{35} homohopanes than the Sockeye oil. Also, the California API-11 has significantly higher concentrations of C_{35} homohopanes ($22S + 22R$) than C_{34} homohopanes ($22S + 22R$), further indicating a highly reducing marine environment of deposition with no available free oxygen.[122] For the Orinoco Bitumen, C_{23} tricyclic terpane is the most abundant, followed by the C_{30} and C_{29} hopane; whereas the Boscan oil demonstrates higher concentrations of C_{29} and C_{30} terpanes than C_{23} terpane. The presence of triplets with different relative distributions is apparent for most heavy oils.

As for distilled petroleum products, refining processes can often remove most high molecular weight (MW) biomarkers from the original crude oil feedstock and only traces of tricyclic to pentacyclic terpanes and steranes are detected in light fuels such as jet fuels and many No. 2 diesels. For the same reasons, these high MW biomarkers are concentrated in heavy residual fuels, lubricating oils and asphalts. For example, many lubricating oils, in general, contain high levels of target terpane and sterane compounds in comparison with most crude oils and petroleum products. The dominance of characteristic pentacyclic C_{31} to C_{35} homohopanes is also often prominent.

Comparison of triterpane (and other) distribution patterns and profiles is widely used for oil correlation and differentiation in environmental forensic studies. For example, on 28 March 2001, three unknown oil samples were received from Montreal for product characterization, correlation and differentiation.[147] These three samples showed nearly identical GC profiles and n-alkane and isoprenoid distribution patterns. However, triterpane biomarker characterization results reveal that (1) samples 1 and 2 had almost identical chromatograms of biomarker terpanes; but sample 3 had a different biomarker distribution than samples 1 and 2: concentrations of C_{23} and C_{24} and the sum of C_{31}–C_{35} homohopanes were significantly lower (39 vs. 145–152 $\mu g\,g^{-1}$ oil for C_{23} and 36 vs. 83–87 $\mu g\,g^{-1}$ oil for C_{24}) and higher (2072 vs. 1358–1376 $\mu g\,g^{-1}$ oil for C_{31}–C_{35}) than those of the corresponding compounds in samples 1 and 2, respectively; (2) the diagnostic ratios of target triterpanes were similar for samples 1 and 2 in addition to the triplet ratios, but sample 3 had significantly different diagnostic ratios of target biomarkers including C_{23}/C_{24}, C_{23}/C_{30}, C_{24}/C_{30}, $C_{30}/(C_{31}$–$C_{35})$ and Ts/Tm. All these observations led to the conclusion that samples 1 and 2 were identical, whereas sample 3 was derived from another source.[147] Similar approaches have been applied to the investigation of oil spill accidents.[1,95,124–128,132,148–155]

Lubricating oil contamination through engine exhaust and through leakage and spillage is ubiquitous in urban environments.[1,156] Bieger et al.[157] reported the use of triterpane biomarker fingerprints of refined oils and motor exhausts to indicate the presence of and trace the origin of diffuse lubricating oil contamination in plankton and sediments around St. John's, Newfoundland, Eastern Canada. Different types of lubricating oils and motor exhausts were found consistently to feature distinct triterpane distributions. Variable inputs of automotive and outboard motor oils were also clearly recognized.

2.9.3 Pentacyclic Triterpanes – Non-hopanoids. In addition to the hopanes, several other pentacyclic triterpanes can occur in petroleum and prove diagnostic in environmental forensic investigations. These diagnostic non-hopanoid triterpanes include gammacerane (Figure 8), oleananes and bicadinanes (Figure 10), which are described in the following paragraphs.

Gammacerane is a C_{30} triterpane containing five six-membered carbon rings (Figure 8). Large amounts of gammacerane in crude oil indicate highly reducing, hypersaline conditions during deposition of the contributing organic matter.[122] Gammacerane has been found in high concentrations (up to 50% in relation to C_{30} αβ-hopane) in the African Angola Cabinda and Gabon crude oils.[158] An example of how the varying amounts of gammacerane – as measured by the gammacerane ratio (Table 8) – can prove diagnostic in spilled oil was presented by Stout and Wang.[6]

Oleananes and their derivatives form the largest group of non-hopanoid triterpenoids.[118] Oleananes occur widely in the plants and are formed in sediments through diagenetic and catagenetic alteration of oleanenes and 3β-functionalized angiosperm triterpenoids.[159] Oleanane has two isomers: 18α(H)-oleanane and 18β(H)-oleanane (Figure 10). The α-type configuration has the highest thermodynamic stability, hence it is the predominant configuration in mature crude oils and rocks. The presence of 18α(H)-oleanane in benthic sediments in Prince William Sound, combined with its absence in Alaska North

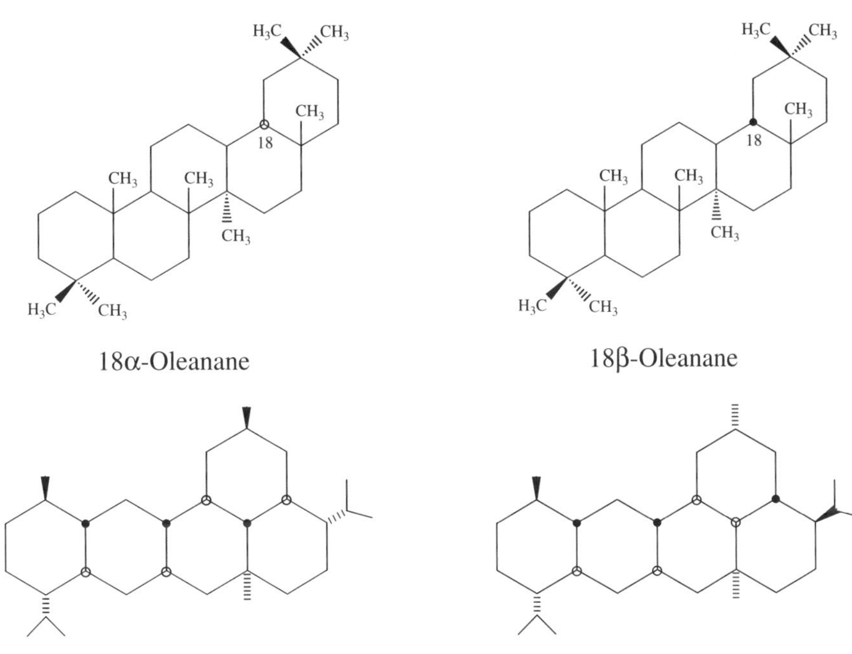

18α-Oleanane

18β-Oleanane

T: trans-trans-trans-bicadinane

W: cis-cis-cis-bicadinane

Figure 10 Molecular structures of selected non-hopanoid, pentacyclic triterpanes whose abundance can be diagnostic for different petroleum sources.

Slope crude and specifically in Exxon Valdez oil and its residue, confirmed
another petrogenic source.[126,149,160] Characterization of 18α(H)-oleanane in
oils from the Anaco area and Maturin subbasin, Venezuela, has been used as an
organic type and age indicator for assessment of Venezuelan petroleum system.[159] Nigerian crude oils from the Niger Delta are characterized by the
presence of oleanane in relatively high concentrations.[158] Recently, Stout
et al.[161] reported the use of chemical fingerprinting to establish the presence of a
spilled Nigerian Bonnie Light-blend crude oil in a residential area following
Hurricane Katrina, Louisiana, in September 2005. Inspection of the GC–MS
m/z 191 extracted ion profiles for the spilled Nigerian blend crude oil (weathered and unweathered) revealed that the concentration of oleanane in the
spilled oil was approximately 30% higher than that of hopane (Figure 11). The
predominance of oleanane in the spilled oil was very unusual among crude oils
worldwide and this fact, in combination with oleananes' resistance to weathering, provided a useful means of (1) distinguishing the spilled crude oil from
other non-crude oils, such as lubricating, hydraulic and transmission oils also

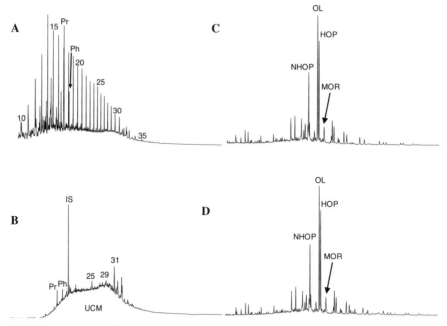

Figure 11   GC–FID of (A) unweathered and (B) weathered Nigerian crude oil blend
spilled in the course of Hurricane Katrina and the corresponding partial
m/z 191 EIPs (C and D, respectively) showing the unusually high concentration of oleanane in the spilled oil. Note the consistency in the triterpane distribution despite weathering and the increased concentration of
biomarkers during weathering. Numbers on peaks, *n*-alkane carbon
number; UCM, unresolved complex mixture; NHOP, 17α(H),21β(H)-
norhopane; HOP, 17α(H),21β(H)-hopane; OL, oleanane; MOR, 17β(H),
21α(H)-moretane; IS, internal standards. Modified from Stout *et al.*[161]

released in the area, (2) evaluating the spatial extent of crude oil contamination and (3) recognizing the presence of small quantities of crude oil when mixed with the allochthonous, natural organic matter (NOM) that dominated many of the sediment/soil samples.[161]

Bicadinanes are C_{30}-pentacyclic biomarker compounds and have three configurations, labeled W (*cis,cis,trans*-bicadinane), T (*trans,trans,trans*-bicadinane) and R (bicadinane; Figure 10). The mass spectra of bicadinanes contain prominent m/z 191 and 217 fragments, while peaks can appear in corresponding chromatograms of both hopanes and steranes. All three bicadinanes (W, T and R forms) elute prior to C_{29} hopane in the m/z 191 chromatogram, but it can be conveniently monitored with little interference using the m/z 412 mass chromatogram. They are believed to originate from the angiosperm Dammar resin and are thus highly specific for resinous input from certain higher plants.[162] Crude oils from the North, Central and South Sumatra basins, Indonesia, have been characterized and the biomarker compositions of the crude oils have been used to classify samples into three types.[163] Type 3 oils were distinguished from Type 1 and Type 2 oils by the prominence of bicadinanes relative to C_{30} hopane, together with high abundances of 18α(H)- and 18β(H)-oleanane on the m/z 412 mass chromatogram.

2.10 Steroids

Steroids are compounds whose structures are based on the tetracyclic androstane ring system. The four rings are designated A, B, C and D, beginning with the ring at lower left and the carbon atoms are numbered beginning with A ring and ending with two 'angular' (axial) methyl groups.[123] The four-ringed saturated steroids are ubiquitous in petroleum and form a class of biomarkers containing 21–30 carbons, including regular steranes, rearranged diasteranes and mono- and triaromatic steranes. Among them, the regular C_{27}–C_{28}–C_{29} homologous sterane series (cholestane, ergostane and stigmastane; Figure 12) are ubiquitous in crude oils and are considered diagnostic in chemical fingerprinting studies because of their high specificity. These sterane homologue series do not contain an integral number of isoprene subunits and therefore only approximate the isoprene rule. Steranes originate from natural occurring four-ringed sterols that are widespread in animals and plants including phytoplankton, zooplankton and vascular plants.[122,164,165]

2.10.1 Normal and Rearranged Steranes. As mentioned above, a series of steranes containing 21–30 carbons, including regular steranes and rearranged steranes, *i.e.* diasteranes, are present in crude oils (Table 9). Steranes are thermally stable and relatively resistant to biodegradation. These properties enhance their utility in environmental forensics investigations. The most straightforward application of steranes in forensic studies is the proportion of C_{27}–C_{28}–C_{29} regular steranes (cholestanes, $C_{27}H_{48}$; ergostanes, $C_{28}H_{50}$; and stigmastanes, $C_{29}H_{52}$) as measured at m/z 218 and 217 (Table 9). The relative distribution of C_{27}–C_{29} αββ sterane (m/z 218) in crude oil reflects the relative proportion of algae

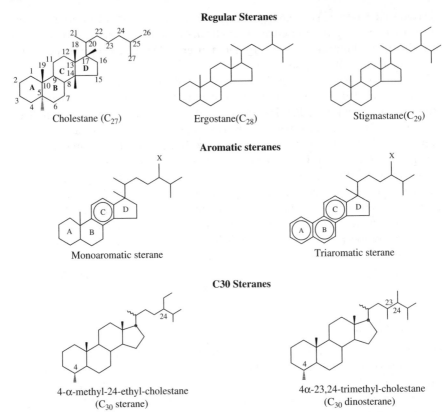

Figure 12   Molecular structures of selected regular steranes, aromatic steroids and C_{30} steranes whose abundance can be diagnostic for different petroleum sources.

versus land plant debris in crude oil's source rock. The characteristic V-shaped C_{27}–C_{28}–C_{29} regular $\alpha\beta\beta$ sterane (m/z 218) distribution indicates high thermal maturity. In general, a dominance of C_{27} over C_{29} steranes specifies marine algae organic matter input, while a predominance of C_{29} steranes over C_{27} steranes may indicate a preferential higher plant input.[122,166]

Different oils often show different sterane distribution patterns. Heavy Orinoco and Boscan oils have somewhat the V-shaped C_{27}–C_{28}–C_{29} regular $\alpha\beta\beta$ sterane distribution. Three California oils (API 11, Platform Elly and Sockeye) have very high concentrations of steranes with a more abundant C_{28} ergostane than C_{27} and C_{29} sterane series. This is also the case for several other heavy California oils including California API–15 and Platform Irene. The ANS, South Louisiana and Troll oils contain higher amounts of diasteranes in addition to C_{21} and C_{22} regular steranes. In contrast, the Arabian Light has much lower concentrations of steranes in total (the total of C_{27}–C_{28}–C_{29} $\alpha\beta\beta$ steranes is only 110 μg g^{-1} oil), but displays significantly higher concentration of C_{29} $\alpha\beta\beta$ steranes than C_{27} $\alpha\beta\beta$ cholestane and C_{28} $\alpha\beta\beta$ ergostane series. As

Table 9 Regular sterane and rearranged steranes (diasteranes) frequently used for forensic oil spill studies.

Peak No.	Compound	Key	Empirical formula	Target ions (m/z)
41	C_{20} 5α(H),14α(H),17α(H)-sterane	S20	$C_{20}H_{34}$	217 and 218
42	C_{21} 5α(H),14β(H),17β(H)-sterane	S21	$C_{21}H_{36}$	217 and 218
43	C_{22} 5α(H),14β(H),17β(H)-sterane	S22	$C_{22}H_{38}$	217 and 218
44	C_{27} 20S-13β(H),17α(H)-diasterane	DIA27S	$C_{27}H_{48}$	217 and 218, 259
45	C_{27} 20R-13β(H),17α(H)-diasterane	DIA27R	$C_{27}H_{48}$	217 and 218, 259
46	C_{27} 20S-13α(H),17β(H)-diasterane	DIA27S2	$C_{27}H_{48}$	217 and 218, 259
47	C_{27} 20R-13α(H),17β(H)-diasterane	DIA27R2	$C_{27}H_{48}$	217 and 218, 259
48	C_{28} 20S-13β(H),17α(H)-diasterane	DIA28S	$C_{28}H_{50}$	217 and 218, 259
49	C_{28} 20R-13β(H),17α(H)-diasterane	DIA28R	$C_{28}H_{50}$	217 and 218, 259
50	C_{29} 20S-13β(H),17α(H)-diasterane	DIA29S	$C_{29}H_{52}$	217 and 218, 259
51	C_{29} 20R-13α(H),17β(H)-diasterane	DIA29R	$C_{29}H_{52}$	217 and 218, 259
52	C_{27} 20S-5α(H),14α(H),17α(H)-cholestane	C27S	$C_{27}H_{48}$	217 and 218
53	C_{27} 20R-5α(H),14β(H),17β(H)-cholestane	C27ββR	$C_{27}H_{48}$	217 and 218
54	C_{27} 20S-5α(H),14β(H),17β(H)-cholestane	C27ββS	$C_{27}H_{48}$	217 and 218
55	C_{27} 20R-5α(H),14α(H),17α(H)-cholestane	C27R	$C_{27}H_{48}$	217 and 218
56	C_{28} 20S-5α(H),14α(H),17α(H)-ergostane	C28S	$C_{28}H_{50}$	217 and 218
57	C_{28} 20R-5α(H),14β(H),17β(H)-ergostane	C28ββR	$C_{28}H_{50}$	217 and 218
58	C_{28} 20S-5α(H),14β(H),17β(H)-ergostane	C28ββS	$C_{28}H_{50}$	217 and 218
59	C_{28} 20R-5α(H),14α(H),17α(H)-ergostane	C28R	$C_{28}H_{50}$	217 and 218
60	C_{29} 20S-5α(H),14α(H),17α(H)-stigmastane	C29S	$C_{29}H_{52}$	217 and 218
61	C_{29} 20R-5α(H),14β(H),17β(H)-stigmastane	C29ββR	$C_{29}H_{52}$	217 and 218
62	C_{29} 20S-5α(H),14β(H),17β(H)-stigmastane	C29ββS	$C_{29}H_{52}$	217 and 218
63	C_{29} 20R-5α(H),14α(H),17α(H)-stigmastane	C29R	$C_{29}H_{52}$	217 and 218
64	C_{30} steranes	C30S	$C_{30}H_{54}$	217 and 218

discussed above, only traces of smaller steranes can be detected in lighter petroleum products, but highly abundant steranes can be found in heavy residual fuels and lubricating oils with the dominance of C_{27}, C_{28} and C_{29} 20S/20R homologues being apparent in many lubricating oils. For example, Extreme Pressure Gear Oil 80W-90 contains high concentrations of C_{27}–C_{30} regular steranes with distribution pattern of $C_{27} < C_{28} < C_{29}$ αββ epimers, whereas 2-Cycle-Engine-Oil contains high amounts of diasteranes.[32]

The distribution and diagnostic ratios of steranes are widely used for oil spill identification and differentiation and some of the common ratios are given in

Table 10. Sterane distributions were useful on a harbor spill that occurred in The Netherlands in 2004. A thick layer of oil (sample 2) was found between a bunker boat and the quay next to the bunker center and it was suspected that something had gone wrong during bunkering of the vessel. Fuel oils from the bunker boat (sample 1) and the bunker center (sample 3) were collected as suspected sources for comparison with the spill sample. A multi-criterion approach was applied to fingerprint and identify these oil samples and to determine the source of the spill.[72] Product-type screening analysis by GC–FID showed that the three samples had very similar GC profiles. Characterization of highly abundant sesquiterpanes by GC–MS indicated sample 1 (bunker boat) being closer to the spill sample 2 than sample 3 (bunker center), evidenced by nearly identical distribution patterns of sesquiterpanes of samples 1 and 2 and also by identical diagnostic ratios of eight sesquiterpane isomers for samples 1 and 2. The source identification was further confirmed by quantitative evaluation of the pentacyclic biomarker terpanes and steranes: (1) traces of terpanes and steranes were detected in the samples, mostly lower MW C_{19}–C_{24} terpanes, diasteranes and C_{27}–C_{29} steranes and no C_{33}–C_{35} pentacyclic hopanes were detected; (2) samples 2 and 1 had nearly identical terpane and sterane distribution patterns; (3) sample 3 showed a distribution pattern different from that of samples 1 and 2, that is, sample 3 had much lower levels of pentacyclic terpanes (C_{29}–C_{32}) and C_{27}–C_{29} steranes than samples 1 and 2; (4) the diagnostic ratios of target hopanes and steranes were similar for samples 1 and 2, whereas the diagnostic ratios of sample 3 were significantly different from either. Clearly, the fingerprinting and quantitation data of biomarker terpanes

Table 10 Diagnostic ratios among steranes that are frequently used for identification, correlation and differentiation of spilled oils (after Wang *et al.*[32]).

Diagnostic ratio	Code
C_{27} 20S-13β(H),17α(H)-diasterane/C_{27} 20R-13β(H), 17α(H)-diasterane	DIA 27S/DIA 27R
Relative distribution of regular C_{27}–C_{28}–C_{29} steranes	C_{27}:C_{28}:C_{29} steranes
C_{27} $\alpha\beta\beta$/C_{29} $\alpha\beta\beta$ steranes (at m/z 218)	C_{27} $\beta\beta(S+R)$/C_{29} $\beta\beta(S+R)$
C_{28} $\alpha\beta\beta$/C_{29} $\alpha\beta\beta$ steranes (at m/z 218)	C_{28} $\beta\beta(S+R)$/$C_{29}\beta\beta(S+R)$
C_{27} $\alpha\beta\beta$/(C_{27} $\alpha\beta\beta$ + C_{28} $\alpha\beta\beta$ + C_{29} $\alpha\beta\beta$) (at m/z 218)	C_{27} $\beta\beta$/(C_{27} + C_{28} + C_{29}) $\beta\beta$
C_{28} $\alpha\beta\beta$/(C_{27} $\alpha\beta\beta$ + C_{28} $\alpha\beta\beta$ + C_{29} $\alpha\beta\beta$) (at m/z 218)	$C_{28}\beta\beta$/(C_{27} + C_{28} + C_{29}) $\beta\beta$
C_{29} $\alpha\beta\beta$/(C_{27} $\alpha\beta\beta$ + C_{28} $\alpha\beta\beta$ + C_{29} $\alpha\beta\beta$) (at m/z 218)	C_{29} $\beta\beta$/(C_{27} + C_{28} + C_{29}) $\beta\beta$
C_{27}, C_{28} and C_{29} $\alpha\alpha\alpha$/$\alpha\beta\beta$ epimers (at m/z 217)	C_{27} $\alpha\alpha$/C_{27} $\beta\beta$
	C_{28} $\alpha\alpha$/C_{28} $\beta\beta$
	C_{29} $\alpha\alpha$/C_{29} $\beta\beta$
C_{27}, C_{28} and C_{29} 20S/(20S + 20R) steranes (at m/z 217)	C_{27} (20S)/C_{27} (20R)
	C_{28} (20S)/C_{28} (20R)
	C_{29} (20S)/C_{29} (20R)
C_{30} sterane index: C_{30}/(C_{27}–C_{30}) steranes	C_{30}/(C_{27}–C_{30}) steranes
Selected diasteranes/regular steranes	
Regular C_{27}–C_{28}–C_{29} steranes/C_{30} $\alpha\beta$-hopanes	C_{27}–C_{28}–C_{29} steranes/H30

and steranes further confirmed the conclusion obtained from the fingerprinting results of sesquiterpanes, that is, sample 1 (bunker boat) is a 'positive match' to the spill sample 2 (spill oil on the water surface), whereas sample 3 (bunker center) is a 'non-match' to the spill.

Stout *et al.*[1] applied the ternary diagram of the relative distribution of C_{27}–C_{29} αββ steranes for a number of tarballs and two candidate source oils. The distribution of C_{27}–C_{28}–C_{29} αββ steranes demonstrated that the highly weathered tarballs, which had washed up on a beach following an oil spill, were from two distinct sources, mostly from candidate source 2. In this case, source 2 was recognized as being derived from local, natural oil seeps, a chronic regional oil source unrelated to the spilled oil (source oil 1).

4α-Methyl-24-ethylcholestanes – often termed C30 steranes (Figure 12) – occur in relatively high abundance in Tertiary source rocks and related oils from China.[167] Almost all of the oils from the eastern Pearl River Mouth Basin (South China Sea) contain significant amounts of 4-methylsteranes.[168] In 2006, an oil spill was reported in the North China Sea and its beaches and after large-scale sample collection and analysis, the North China Sea Monitoring Center found that the spill was from an oil tanker carrying oil from the South China Sea,[169] rather from the oil spills from oil platforms located in the North China Sea. The fingerprinting evidence to support this conclusion included: (1) the spill oil (lightly weathered) and the source candidate showed almost identical GC–FID profiles with the same C_{17}:pristine and C_{18}:phytane ratios; (2) both the spill and source candidate samples exhibited nearly identical and very unique distributions of steranes with the C_{27}–C_{28}–C_{29} αββ steranes being almost equally distributed and with C_{30} 4-methylsteranes being unusually rich and even more abundant than C_{29} αββ sterane homologues; and (3) oleanane and gammacerane with nearly the same concentrations were detected in both the spill and source candidate samples.

2.10.2 Aromatic Steroids. Aromatic steroids, which include C-ring mono-aromatic (MA) steroids and ABC-ring triaromatic (TA) steroids, are another group of biomarker compounds found in the oil aromatic hydrocarbon fraction (Figure 12). The C-ring MA steroids may be derived exclusively from sterols with the side-chain double bond during early diagenesis, while the ABC-ring TA steroids are formed from aromatization of C-ring MA steroids involving the loss of a methyl group at the A/B ring junction.[170,171] The MA steroids are composed mainly of a series of 20*R* and 20*S* 5α- and 5β-cholestanes, ergostanes and stigmastanes and also rearranged ring-C 20*S* and 20*R* MA-diasteranes. TA steroids are composed mainly of C_{26}–C_{27}–C_{28} homologous plus C_{20}–C_{22} TA steroids. Examples of MA and TA steroids are shown in Figure 12.

Aromatic steroids are highly resistant to biodegradation and can provide useful diagnostic information on the nature of the precursor organic matter type and thermal history of crude oil and, thereby, information for oil-to-oil correlation and oil source tracking. Aromatic steroids are monitored using m/z 231 and 253 for TA and MA steroids, respectively (Table 1). Generally, TA steroids are much more abundant than MA steroids for all oils studied.

Different oils and refined products show apparent differences in the relative distributions and absolute concentrations of TA and MA steroids. In many lighter oils such as Cook Inlet, Federated, West Texas and Scotia Light, only trace MA steroids are detected. This implies that TA steroids are more valuable marker compounds than MA steroids for environmental forensic investigations. Unlike most Canadian diesels, the Korean Diesel No. 2 still contains a relatively large quantity of high carbon number TA steroids. Similarly, lubricating oils contain no or only trace levels of MA-sterane compounds. Synthetic lubricants should not contain any TA- or MA-sterane compound if they are purely chemically synthesized. Barakat *et al.*[172] have reported a case study in which oil residues were correlated to a fresh crude oil sample of Egyptian Western Desert-sourced oil by fingerprinting MA and TA steroids and by determination and comparison of molecular ratios of the target MA- and TA-sterane compounds, including $C_{28}TA$ $20R/C_{28}TA$ $20S$, $C_{27}TA$ $20R/C_{28}TA$ $20R$ and $C_{28}TA$ $20S/(C_{26}TA$ $20R + C_{27}TA$ $20S)$.

3 Conclusions

The complex chemical makeup of petroleum – which can contain tens of thousands of individual hydrocarbons and non-hydrocarbons – provides an opportunity to 'chemically fingerprint' petroleum contamination in the environment and thereby assess its relationship among impacted samples and to known or suspected sources.

No one group of hydrocarbons or non-hydrocarbons is universally 'diagnostic' for the chemical fingerprinting of petroleum in the environment. However, when multiple compound groups – discussed herein – are appropriately applied to a specific petroleum-related question, they afford a powerful assemblage for the characterization of and comparison among petroleum samples in the environment. The common characteristics of diagnostic compounds include:

1. the potential for a high degree of specificity among different oils or fuels,
2. relative widespread occurrence in petroleum,
3. ability to be measured precisely within a complex mixture and in multiple matrices, and perhaps most importantly,
4. relative resistance to environmental weathering.

Because most diagnostic compounds are not regulated chemicals, their analysis requires modifications to the standard analytical methods used for regulatory purposes. The distributions and abundance of diagnostic compounds should be evaluated using ratios or other metrics reliant upon quantitative (*i.e.* absolute concentrations) or semiquantitative (*e.g.* peak areas or heights) data usually acquired using GC–MS. These ratios or other metrics can be compared using numerous graphical, statistical or numerical correlation tools. However, because diagnostic compounds might represent only a small fraction of the total petroleum, the utility of qualitative fingerprinting (chromatographic comparison) used in conjunction with quantitative fingerprinting of diagnostic compounds cannot be ignored.

References

1. S. A. Stout, A. D. Uhler, K. J. McCarthy and S. Emsbo-Mattingly, in *Introduction to Environmental Forensics*, ed. B. L. Murphy and R. D. Morrison, Academic Press, London, 2002, pp. 137–260.
2. Z. Wang, M. Fingas, C. Yang and J. H. Christensen, in *Environmental Forensics Contaminant Specific Guide*, ed. R. D. Morrison and B. L. Murphy, Academic Press, Boston, 2006, pp. 339–407.
3. W. K. Seifert, J. M. Moldowan and R. W. Jones, in *Tenth World Congress*, Heyden, London, 1979, pp. 425–440.
4. Z. Wang, S. A. Stout and M. Fingas, *Environ. Forensics*, 2006, **7**, 105–146.
5. R. C. Prince, E. H. Owens and G. A. Sergy, *Mar. Pollut. Bull.*, 2002, **44**, 1236–1242.
6. S. A. Stout and Z. Wang, in *Oil Spill Environmental Forensics*, ed. Z. Wang and S. A. Stout, Academic Press, Boston, 2007, pp. 1–53.
7. L. Mansuy, R. P. Philp and J. Allen, *Environ. Sci. Technol.*, 1997, **31**, 3417–3425.
8. G. S. Frysinger and R. B. Gaines, *J. High Resolut. Chromatogr.*, 1999, **22**, 251–255.
9. R. Rodgers, E. N. Blumer, M. A. Freitas and A. G. Marshall, *Environ. Sci. Technol.*, 2000, **34**, 1671–1678.
10. B. P. Tissot and D. H. Welte, *Petroleum Formation and Occurrence*, Springer-Verlag, New York, 1984.
11. K. E. Peters and M. G. Fowler, *Org. Geochem.*, 2002, **33**, 5–36.
12. K. E. Peters, G. L. Scheuerman, C. Y. Lee, J. M. Moldowan, R. N. Reynolds and M. M. Pena, *Energy Fuels*, 1992, **6**, 560–577.
13. B. Hamilton and R. J. Falkiner, in *Fuels and Lubricants Handbook: Technology, Properties, Performance and Testing*, ed. G. E. Totten, ASTM Manual Series, MNL37WCD, ASTM International, West Conshohocken, PA, 2003, pp. 61–88.
14. L. B. Christensen and T. H. Larsen, *Ground Water Monitoring and Remediation*, 1993, Fall, 142–149.
15. I. R. Kaplan, Y. Galperin, H. Alimi, R. P. Lee and S.-T. Lu, *Ground Water Monitoring and Remediation*, 1996, Fall, 113–124.
16. S. A. Stout and A. D. Uhler, *Environ. Claims J.*, 2003, **15**, 241–259.
17. American Society for Testing and Materials, *ASTM D-3328-90*, ASTM, Philadelphia, PA, 1990.

18. American Society for Testing and Materials, *ASTM D-5739-00*, ASTM International, West Conshohocken, PA, 2000.
19. P. Daling, L. Faksness, A. Hansen and S. Stout, Source ID of waterborne oil spills using standardized methodology, *Environ. Forensics*, 2003, **3**, 263–278.
20. J. Christensen, G. Tomasi and A. B. Hansen, *Environ. Sci. Technol.*, 2005, **39**, 255–260.
21. J. H. Christensen, J. Mortensen, A. B. Hansen and O. Andersen, *J. Chromatogr. A*, 2005, **1062**, 113–123.
22. A. B. Hansen, P. S. Daling, L. G. Faksness, K. R. Sorheim, P. Keinhuis and R. Duus, in *Oil Spill Environmental Forensics*, ed. Z. Wang and S.A. Stout, Academic Press, Boston, 2007, pp. 229–256.
23. G. S. Douglas and A. D. Uhler, *Environ. Test. Anal.*, 1993, **5**, 46–53.
24. G. S. Douglas, S. D. Emsbo-Mattingly, S. A. Stout, A. D. Uhler and K. J. McCarthy, in *Introduction to Environmental Forensics*, 2nd edn., ed. B. L. Murphy and R. D. Morrison, Academic Press, Boston, 2007, pp. 312–454.
25. J. H. Christensen, A. B. Hansen, G. Tomasi, J. Mortensen and O. Andersen, *Environ. Sci. Technol.*, 2004, **38**, 2912–2918.
26. G. W. Johnson and R. Ehrlich, *Environ. Forensics*, 2002, **3**, 59–79.
27. Z. Wang and J. H. Christensen, in *Environmental Forensics Contaminant Specific Guide*, ed. R. D. Morrison and B. L. Murphy, Academic Press, Boston, 2006, pp. 409–464.
28. T. D. Gauthier and M. Hawley, in *Introduction to Environmental Forensics*, 2nd edn., ed. B. L. Murphy and R. D. Morrison, Academic Press, Boston, 2007, pp. 129–183.
29. W. A. Burns, S. M. Mudge, A. E. Bence, P. D. Boehm, S. Brown John, D. S. Page and K. R. Parker, *Environ. Sci. Technol.*, 2006, **40**, 6561–6567.
30. G. S. Douglas, S. A. Stout, A. D. Uhler, K. J. McCarthy and S. D. Emsbo-Mattingly, in *Oil Spill Environmental Forensics*, ed. Z. Wang and S.A. Stout, Academic Press, Boston, 2007, pp. 257–292.
31. J. H. Christensen and G. Tomasi, in *Oil Spill Environmental Forensics*, ed. Z. Wang and S. A. Stout, Academic Press, Boston, 2007, pp. 293–325.
32. Z. D. Wang, C. Yang, M. Fingas, B. Hollebone, U. H. Yim and J. R. Oh, in *Oil Spill Environmental Forensics*, ed. Z. D. Wang and S. A. Stout, Academic Press, Boston, 2007, pp. 73–146.
33. L. R. Durrett, L. M. Taylor, D. F. Wantland and I. Dvoretzky, *Anal. Chem.*, 1963, **35**, 637–641.
34. J. G. Speight, *The Chemistry and Technology of Petroleum*, Marcel Dekker, New York, 1991.
35. T. Hutson and G. E. Hays, in *Industrial and Laboratory Alkylations*, ed. L. F. Albright and A. R. Goldsby, ACS Symposium Series No. 55, American Chemical Society, Washington, DC, 1977, pp. 27–55.
36. H. Pines, *The Chemistry of Catalytic Hydrocarbon Conversions*, Academic Press, New York, 1981.

37. P. W. Beall, S. A. Stout, G. S. Douglas and A. D. Uhler, *Environ. Claims J.*, 2002, **14**, 487–506.
38. S. A. Stout, G. S. Douglas and A. D. Uhler, in *Environmental Forensics – Contaminant Specific Approach*, ed. R. Morrison and B. Murphy, Academic Press, Boston, 2006. pp. 463–531.
39. A. Marchetti, J. Daniels and D. Layton, *Potential Ground Surf. Water Impacts*, 1999, **4**, 1–20.
40. L. M. Gibbs, *Gasoline Additives – When and Why*, SAE Technical Paper Series, Int'l. Fuels and Lubricants Mtg., Tulsa, OK, 1990.
41. I. R. Kaplan, *Environ. Forensics*, 2003, **4**, 95–141.
42. I. R. Kaplan, Y. Galperin, S.-T. Lu and R.-P. Lee, *Org. Geochem.*, 1997, **27**, 289–317.
43. M. D. Johnson and R. D. Morrison, *Environ. Protect.*, 1996, September, 37–39.
44. P. T. Mulroy and L.-T. Ou, *Environ. Toxicol. Chem.*, 1998, **17**, 777–782.
45. American Society for Testing and Materials, *D3695-Standard Test Method for volatile alcohols in water by direct aqueous injections gas chromatography*, ASTM International, West Conshahocken, PA, 2001.
46. S. E. Powers, D. Rice, B. Dooher and P. J. J. Alvarez, *Environ. Sci. Technol./News* 1 January 1, 2001, 24A–30A.
47. J. M. Davidson and D. N. Creek, *Environ. Forensics*, 1999, **1**, 57–67.
48. A. D. Uhler, S. A. Stout, R. M. Uhler, S. D. Emsbo-Mattingly and K. J. McCarthy, *Environ. Forensics*, 2001, **2**, 17–19.
49. R. C. Fort, *Adamantane: The Chemistry of Diamond Molecules*, Marcel Dekker, New York, 1976.
50. W. S. Wingert, *Fuel*, 1992, **71**, 37–43.
51. J. E. Dahl, J. M. Moldowan, K. E. Peters, G. E. Claypool, M. A. Rooney and G. E. Michael, *Nature*, 1999, **399**, 54–57.
52. M. Radke, H. Willsch and M. Teichmuller, *Org. Geochem.*, 1990, **15**, 539–563.
53. K. Grice, R. Alexander and R. I. Kagi, *Org. Geochem.*, 2000, **31**, 67–73.
54. Z. Wang, C. Yang, B. Hollebone and M. Fingas, *Environ. Sci. Technol.*, 2006, **40**, 5636–5646.
55. S. A. Stout and G. S. Douglas, *Environ. Forensics*, 2004, **5**, 225–235.
56. J. Chen, J. Fu, G. Sheng, D. Liu and J. Zhang, *Org. Geochem.*, 1996, **25**, 179–190.
57. L. Jinggui, P. Philp and C. Mingzhong, *Org. Geochem.*, 2000, **31**, 267–272.
58. L. K. Schulz, A. A. Williams, E. Rein and A. S. Steen, *Org. Geochem.*, 2001, **32**, 365–375.
59. G. A. Caravaggio, J. P. Charland, P. MacDonald and L. Graham, *Environ. Sci. Technol.*, 2007, **41**, 3697–3701.
60. G. Eglinton and R. J. Hamilton, *Science*, 1967, **156**, 1322–1335.
61. E. E. Bray and E. D. Evans, *Geochim. Cosmochim. Acta*, 1961, **22**, 2–15.
62. I. A. Leal-Granadillo, I. J. Alonso and A. Sanz-Medel, *J. Environ. Monit.*, 2000, **2**, 218–222.

63. C. Song, *Introduction to the Chemistry of Diesel Fuels*, Taylor and Francis, New York, 2000.
64. M. J. Wade, *Environ. Forensics*, 2005, **6**, 187–196.
65. S. A. Stout, A. D. Uhler and K. J. McCarthy, *Environ. Forensics*, 2006, **7**, 267–282.
66. K. E. Peters, C. C. Walters and J. M. Moldowan, *The Biomarker Guide*, 2nd edn., Cambridge University Press, Cambridge, 2005.
67. H. Borwitzky and G. Schomburg, *J. Chromatogr. A*, 1982, **240**, 307–318.
68. C. P. McIntyre, P. Harvey, S. H. Ferguson, A. M. Wressnig, H. Volk, S. C. George and I. Snape, *Environ. Sci. Technol.*, 2007, **41**, 2452–2458.
69. J. G. Bendoraitis, *Advances in Org. Geochemistry-1973*, Editions Technip, Paris, 1974, pp. 209–224.
70. R. Alexander, R. I. Kagi, R. A. Noble and J. K. Volkman, *Org. Geochem.*, 1984, **6**, 63–70.
71. S. A. Stout, A. D. Uhler and K. J. McCarthy, *Environ. Forensics*, 2005, **6**, 241–252.
72. Z. Wang, C. Yang, M. Fingas, Hollebone, A. B. Peng, A. B. Hansen and J. H. Christensen, *Environ. Sci. Technol.*, 2005, **39**, 8700–8707.
73. R. B. Gaines, G. J. Hall, G. S. Frysinger, W. R. Gronlund and K. L. Juaire, *Environ. Forensics*, 2006, **7**, 77–87.
74. J. A. Williams, M. Bjoroy, D. L. Dolcater and J. C. Winters, *Org. Geochem.*, 1985, **10**, 451–461.
75. I. R. Kaplan and Y. Galperin, *Contaminated Soils*, 1997, **2**, 65–78.
76. F. D. Hostettler and K. A. Kvenvolden, *Environ. Forensics*, 2002, **3**, 293–301.
77. S. A. Stout and A. D. Uhler, *Environ. Forensics*, 2006, **7**, 283–287.
78. B. R. T. Simoneit*Cyclic Terpenoids of the Geosphere*, 1986, 43–99 in *Biological Markers in the Sedimentary Record*, ed. R. B. Johns, Elsevier, New York, 1986, pp. 43–99.
79. S. Wakeham, C. Schaffner and W. Giger, *Geochim. Cosmochim. Acta*, 1980, **44**, 415–429.
80. L.-G. Faksness, P. S. Daling and A. B. Hansen, *Environ. Forensics*, 2002, **3**, 279–292.
81. P. D. Boehm, in *Environmental Forensics Contaminant Specific Guide*, ed. R.D. Morrison and B.L. Murphy, Academic Press, Boston, 2006, pp. 313–337.
82. W. Giger and M. Blumer, *Anal. Chem.*, 1974, **46**, 1663–1671.
83. W. W. Youngblood and M. Blumer, *Geochim. Cosmochim. Acta*, 1975, **39**, 1303–1314.
84. M. L. Lee, G. P. Prado, J. B. Howard and R. A. Hites, *Biomed. Mass Spectrom.*, 1977, **4**, 182–186.
85. R. E. Laflamme and R. A. Hites, *Geochim. Cosmochim. Acta*, 1978, **42**, 289–303.
86. NOAA, Technical Memorandum NOS/ORCA 71, *National Oceanic and Atmospheric Administration*, 1993, Silver Springs, MD.

87. G. S. Douglas, W. A. Burns, A. E. Bence, D. S. Page and P. D. Boehm, *Environ. Sci. Technol.*, 2004, **38**, 3958–3964.
88. S. A. Stout, A. D. Uhler and S. D. Emsbo-Mattingly, *Environ. Sci. Technol.*, 2004, **38**, 2987–2994.
89. D. L. Elmendorf, C. E. Haith, G. S. Douglas and R. C. Prince, in *Bioremediation of Chlorinated and Polycyclic Aromatic Hydrocarbon Compounds*, ed. R. L. Hinchee, A. Leeson, L. Semprini and S. K. Ong, Lewis, Boca Raton, FL, 1994, pp. 188–202.
90. P. D. Boehm and J. W. Farrington, *Environ. Sci. Technol.*, 1984, **18**, 840–841.
91. M. Blumer and W. W. Youngblood, *Science*, 1975, **188**, 53–55.
92. A. Bjøeseth, in *Handbook of Polycyclic Aromatic Hydrocarbons*, ed. A. Bjøeseth and T. Ramdahl, Marcel Dekker, New York, 1985, pp. 1–20.
93. K. T. Benlahcen, A. Chaoui, H. Budzinski, J. Bellocq and P. Garrigues, *Mar. Pollut. Bull.*, 1997, **34**, 298–305.
94. M. A. Sicre, J. C. Marty, A. Salion, Aparicio, J. Grimalt and J. Albaiges, *Atmos. Environ.*, 1987, **21**, 2247–2259.
95. Z. D. Wang, M. Fingas, Y. Y. Shu, L. Sigouin, M. Landriault and P. Lambert, *Environ. Sci. Technol.*, 1999, **33**, 3100–3109.
96. M. G. Meniconi, I. T. Gabardo, M. E. Carneiro, S. M. Barbanti, G. C. Silva and C. G. Massone, *Environ. Forensics*, 2002, **3**, 303–322.
97. I. Tolosa, S. de Mora, M. Sheikholeslami, J. Villeneuve, J. Bartocci and C. Cattini, *Mar. Pollut. Bull.*, 2004, **48**, 4–60.
98. Z. D. Wang, M. Fingas and P. Lambert, *J. Chromatogr.*, 2004, **1038**, 201–214.
99. Z. D. Wang, K. Li, P. Lambert and C. Yang, *J. Chromatogr. A*, 2007, **1139**, 14–26.
100. J. C. Colombo, E. Pelletier, C. Brochu and M. Khalil, *Environ. Sci. Technol.*, 1989, **23**, 888–894.
101. W. E. Pereira, F. D. Hostettler, S. N. Luoma, A. van Geen, C. C. Fuller and R. J. Anima, *Mar. Chem.*, 1999, **64**, 99–113.
102. R. M. Dickhut, E. A. Canuel, K. E. Gustafson, K. Liu, K. M. Arzayus, S. E. Walker, G. Edgecombe, M. O. Gaylor and E. H. MacDonald, *Environ. Sci. Technol.*, 2000, **34**, 4635–4640.
103. M. B. Yunker, R. W. Macdonald, R. Vingarzan, R. H. Mitchell, D. Goyette and S. Sylvestre, *Org. Geochem.*, 2002, **33**, 489–515.
104. S. A. Stout, A. D. Uhler and S. D. Emsbo-Mattingly, *Soil and Sed. Contam.*, 2003, **12**, 815–834.
105. S. E. Walker, R. M. Dickhut, C. Chisholm-Brause, S. Sylva and C. M. Reddy, *Org. Geochem.*, 2005, **36**, 619–632.
106. H. Budzinski, I. Jones, J. Bellocq, C. Pierard and P. Garrigues, *Mar. Chem.*, 1997, **58**, 85–97.
107. M. Radke, D. H. Welte and H. Willsch, *Org. Geochem.*, 1986, **10**, 51–63.
108. M. Radke, H. Willsch and D. Leythaeuser, *Geochim. Cosmochim. Acta*, 1982, **46**, 1831–1848.

109. M. Radke, Application of aromatic compounds as maturity indicators in source rocks and crude oils, *Marine Petrol Geol.*, 1988, **5**, 224–236.

110. Z. Wang, M. Fingas, S. Blenkinsopp, G. Sergy, M. Landriault, L. Siqouin, J. Foght, K. Semple and D. W. S. Westlake, *J. Chromatogr. A*, 1998, **809**, 89–107.

111. A. Stumpf, K. Tolvay and M. Juhasz, *J. Chromatogr. A*, 1998, **819**, 67–74.

112. R. Coulombe, *J. Forensic Sci.*, 1995, **40**, 867–873.

113. Z. Wang and M. Fingas, *Environ. Sci. Technol.*, 1995, **29**, 2842–2849.

114. Z. Wang and M. Fingas, *Mar. Pollut. Bull.*, 2003, **47**, 423–452.

115. G. S. Douglas, A. E. Bence, R. C. Prince, S. J. McMillen and E. L. Butler, *Environ. Sci. Technol.*, 1996, **30**, 2332–2339.

116. T. Schade, B. Roberz and J. T. Anderson, *Polycycl. Arom. Compd.*, 2002, **22**, 311–320.

117. A. Hegazi and J. T. Andersson, in *Oil Spill Environmental Forensics*, ed. Z. Wang and S.A. Stout, Academic Press, Boston, 2007, pp. 147–168.

118. J. D. Connolly and R. A. Hill, *Dictionary of Terpenoids*, Chapman and Hall, London, 1991.

119. A. D. Uhler, S. A. Stout and G. S. Douglas, in *Oil Spill Environmental Forensics*, ed. Z. Wang and S. A. Stout, Academic Press, Boston, 2007, pp. 327–348.

120. K. E. Peters and J. M. Moldowan, *Org. Geochem.*, 1991, **17**, 47–61.

121. T. C. Sauer and P. D. Boehm, *Technical Report Series 95-032*, Marine Spill Response Corp., Washington, DC, 1995.

122. K. E. Peters and J. W. Moldowan, *The Biomarker Guide: Interpreting Molecular Fossils in Petroleum and Ancient Sediments*, Prentice Hall, Englewood Cliffs, NJ, 1993.

123. L. G. Wade Jr., *Organic Chemistry*, 5th edn., Prentice Hall, Upper Saddle River, NJ, 2003.

124. S. A. Stout, A. D. Uhler and K. J. McCarthy, *Environ. Forensics*, 2001, **2**, 87–98.

125. P. D. Boehm, G. S. Douglas, W. A. Burns, P. J. Mankiewicz, D. S. Page and A. E. Bence, *Mar. Pollut. Bull.*, 1997, **34**, 599–613.

126. A. E. Bence, K. A. Kvenvolden and M. C. Kennicutt II, *Org. Geochem.*, 1996, **24**, 7–42.

127. J. K. Volkman, A. T. Revil and A. P. Murray, in *Molecular Markers in Environmental Geochemistry*, ed. R. P. Eganhouse, American Chemical Society, Washinton, DC, 1997, pp. 83–99.

128. K. A. Kvenvolden, F. D. Hostettler, R. W. Rosenbauer, T. D. Lorenson, W. T. Castle and S. Sugarman, *Mar. Geol.*, 2002, **181**, 101–113.

129. F. D. Hostettler, W. E. Pereira, K. A. Kvenvolden, A. Green, S. N. Luoma, C. C. Fuller and R. Anima, *Mar. Chem.*, 1999, **64**, 115–127.

130. M. P. Zakaria, A. Horinouchi, S. Tsutsumi, H. Takada, S. Tanabe and A. Ismail, *Environ. Sci. Technol.*, 2000, **34**, 1189–1196.

131. Z. D. Wang, M. Fingas and K. Li, *J. Chromatogr. Sci.*, 1994, **32**, 361–366 (Part I) and 367–382 (Part II).

132. Z. D. Wang, M. Fingas and G. Sergy, *Environ. Sci. Technol.*, 1994, **28**, 1733–1746.
133. Z. D. Wang, M. Fingas, M. Landriault, L. Sigouin, S. Grenon and D. Zhang, *Environ. Technol.*, 1999, **20**, 851–862.
134. Z. D. Wang, M. Fingas and D. Page, *J. Chromatogr.*, 1999, **843**, 369–411.
135. S. A. Stout, A. D. Uhler, T. G. Naymik and K. J. McCarthy, *Environ. Sci. Technol.*, 1998, **32**, 260A–264A.
136. J. K. Volkman, D. G. Holdsworth, G. P. Neill and H. J. Bavor Jr., *Sci. Total Environ.*, 1992, **112**, 203–219.
137. S. M. Barbanti, Personal communication, 2004.
138. W. K. Seifert and J. M. Moldowan, *Geochim. Cosmochim. Acta.*, 1978, **42**, 77–95.
139. F. R. Aquino Neto, J. M. Trendel, A. Restle, J. Connan, and P. A. Albrecht, in *Advances in Organic Geochemistry 1981*, ed. M. Bjoroy, C. Albrecht and C. Cornford, Wiley, New York, 1983, pp. 659–676.
140. J. M. Trendel, A. Restle, J. Connan and P. A. Albrecht, *J. Chem. Soc., Chem. Commun.*, 1982, 304–306.
141. J. G. Palacas, D. E. Anders and J. D. King, in *Petroleum Geochemistry and Source Rock Potential of Carbonate Rocks*, ed. J. G. Palacas, Studies in Geology, No. 18, American Association of Petroleum Geologists, Houston, TX, 1984, pp. 71–96.
142. K. A. Kvenvolden, J. B. Rapp, J. H. Bourell, in *Alaska North Slope Oil/Rock Correlation Study*, ed. L. B. Magoon and G. E. Claypool, Studies in Geology, No. 20, American Association of Petroleum Geologists, 1985, pp. 593–617.
143. J. K. Volkman, R. Alexander, R. I. Kagi and G. W. Woodhouse, *Geochim. Cosmochim. Acta.*, 1983, **47**, 785–794.
144. J. K. Volkman, R. Alexander, R. I. Kagi and J. Rüllkøtter, *Geochim. Cosmochim. Acta.*, 1983, **47**, 1033–1040.
145. O. Grahl-Nielsen and T. Lygre, *Mar. Pollut. Bull.*, 1990, **21**, 176–183.
146. M. R. Mello, E. A. M. Koutsoukos, M. B. Hart, S. C. Brassell and J. R. Maxwell, *Org. Geochem.*, 1990, **14**, 529–542.
147. Z. D. Wang, M. Fingas and L. Sigouin, *Environ. Forensics*, 2002, **3**, 251–262.
148. A.O. Barakat, A. R. Mostafa, J. Rullkotter and A. R. Hegazi, *Mar. Pollut. Bull.*, 1999, **38**, 535–544.
149. K. A. Kvenvolden, F. D. Hostettler, J. B. Rapp and P. R. Carlson, *Mar. Pollut. Bull.*, 1993, **26**, 24–29.
150. D. S. Page, J. D. Foster, P. M. Fickett and E. S. Gilfillan, *Mar. Pollut. Bull.*, 1988, **3**, 107–115.
151. Z. D. Wang, M. Fingas and G. Sergy, *Environ. Sci. Technol.*, 1995, **29**, 2622–2631.
152. Z. D. Wang, M. Fingas, M. Landriault, L. Sigouin, Y. Feng and J. Mullin, *J. Chromatogr. A*, 1997, **775**, 251–265.

153. Z. D. Wang, M. Fingas, M. Landriault, L. Sigouin, B. Castel, D. Hostetter, D. Zhang and B. Spencer, *J. High Resolut. Chromatogr. A.*, 1998, **21**, 383–395.
154. Z. D. Wang, M. Fingas and L. Sigouin, *J. Chromatogr. A*, 2001, **909**, 155–169.
155. M. P. Zakaria, T. Okuda and H. Takada, *Mar. Pollut. Bull.*, 2001, **12**, 1357–1366.
156. I. R. Kaplan, S. Lu, H. M. Alomi and J. MacMurphey, *Environ. Forensics*, 2001, **2**, 231–248.
157. T. Bieger, J. Helou and T. A. Abrajano Jr, *Mar. Pollut. Bull.*, 1996, **32**, 270–274.
158. G. Dahlmann, *Berichte des BSH 31*, 2003.
159. A. Alberdi and L. Lopez, *J. South Am. Earth Sci.*, 2000, **13**, 751–759.
160. D. S. Page, P. D. Boehm, G. S. Douglas, A. E. Bence, W. A. Burns and P. J. Mankiewicz, *Environ. Toxicol. Chem.*, 1996, **15**, 1266–1281.
161. S. A. Stout, B. Liu, G. C. Millner, D. Hamlin and E. Healey, *Environ. Sci. Technol.*, 2007, **41**, 7742–7751.
162. B. G. K. Van Aarssen, H.C. Cox, P. Hoogendoorn and J.W. deLeeuw, *Geochim. Cosmochim. Acta.*, 1990, **54**, 3021–3031.
163. I. B. Sosrowidjojo, R. Alexander and R. I. Kagi, *Org. Geochem.*, 1994, **21**, 303–312.
164. A. Z. Mackenzie, S. C. Brassell, G. Eglinton and J. R. Maxwell, *Science*, 1982, **217**, 491–504.
165. J. W. de Leeuw, H. C. Cox, G. Van Graas, F. W. Van de Meer, T. M. Peakman, J. M. A. Baas and V. Van de Graaf, *Geochim. Cosmochim. Acta.*, 1989, **53**, 903–909.
166. K. Gürgey, *Org. Geochem.*, 2002, **33**, 723–741.
167. J. Fu, C. Pei, G. Sheng and D. Liu, *J. Southeast Asian Earth Sci.*, 1992, **7**, 271–272.
168. S. Zhang, D. Liang, Z. Gong, K. Wu, M. Li, F. Song, Z. Song, D. Zhang and P. Wang, *Org. Geochem.*, 2003, **34**, 971–991.
169. P. Sun, Personal communication, 2007.
170. J. Riolo, G. Hussler, P. Albrecht and J. Connan, *Org. Geochem.*, 1986, **10**, 981–990.
171. J. M. Moldowan and F. J. Fago, *Geochim. Cosmochim. Acta*, 1986, **50**, 343–351.
172. A. O. Barakat, Y. Qian, M. Kim and M. C. Kennicutt II, *Environ. Forensics*, 2002, **3**, 219–226.

Perchlorate – Is Nature the Main Manufacturer?

IOANA G. PETRISOR AND JAMES T. WELLS

The significant problems we face cannot be solved at the same level of thinking we were at when we created them.

Albert Einstein

1 Introduction

1.1 Changing Perspectives

Changes define the new millennium. Changes are essential to progress. If changes are the engine of progress, then understanding the changes is the key to go forward. Furthermore, foreseeing the changes is the guarantee for success. In environmental forensics, recognizing emerging contamination issues holds the key. To do so, we need to raise our awareness related to any change that can result into an emerging issue.

Perchlorate is a good example of an old contaminant (manufactured and used for more than 100 years) that recently raised emerging issues. In just a few years, our perspective of perchlorate has totally changed: from a useful chemical to an environmental pollutant, and also from what was believed to be purely an anthropogenic to a possibly abundant naturally occurring compound. Specifically, what has been a useful chemical (with many practical applications) and even a medicine (for hyperthyroidism) became an environmental threat due to its widespread occurrence, persistence and capacity to interfere with thyroid gland function. Furthermore, perchlorate is unreactive and refractory to common remediation methods. From a forensic perspective, however, the challenge is not so much related to perchlorate occurrences, toxicity and refractory to degradation, but rather to its unexpected increased detections in the environment and food, of which more and more appear to be due to natural causes. The challenge consists of being able to discern natural

Issues in Environmental Science and Technology, No. 26
Environmental Forensics
Edited by RE Hester and RM Harrison

occurrences in the environment that could also be affected by a variety of anthropogenic activities. Most importantly, it is essential to realize that natural perchlorate formation could be an important source even when anthropogenic sources are obvious.

Today, it is well recognized that perchlorate is naturally formed and present in a series of materials, of which perhaps the best known and characterized is Chilean nitrate deposit from the Atacama Desert.[1,2] Additionally, the appearance of perchlorate in natural mineralogical deposits[3] raises an important question: how did perchlorate get there? In other words, what is the process through which perchlorate is naturally formed? Moreover, if this can happen in the arid area of Chile, why not also somewhere else, especially under similar arid conditions? Are there other natural formation mechanisms that may account for the presence of perchlorate in non-arid areas too? All these questions are of tremendous importance in forensic studies. This is because in order to discern and allocate the source of perchlorate, one needs to consider any possible source first. For perchlorate the possibilities of natural formation, at least in theory, seem endless.

A lot has been written about perchlorate (including books and research and review articles). It is not our intention to repeat or provide a comprehensive review of the existing wealth of information. This chapter has been written from a different, forensic perspective. First, we intend to raise the awareness of the natural formation of perchlorate, discussing both confirmed and highly possible formation scenarios. Second, the main characteristics of perchlorate will be linked to its fate and transport and also its many uses. Finally, we provide a practical forensic guide aiming to identify the investigative methods available for perchlorate source tracking and age dating, illustrated, where available, by references to successful forensic applications. The chapter will emphasize those methods that could be used to delineate natural formation from anthropogenic. In summary, while providing a useful and up-to-date guide for forensic investigations, this chapter also raises many questions and aims to help identify new areas in need of research related to this emerging contaminant. We hope that this chapter will boost forensic studies on perchlorate occurrences along with interest in evaluating natural perchlorate.

1.2   The Perchlorate Legacy – Emergence of a Long-used Contaminant

Following technological changes, emerging issues are blooming: new contaminants are manufactured and possibly released into the environment. Perchlorate is undoubtedly an emerging contaminant. This does not relate to its usage period, since perchlorate has been used for many years. The emerging classification of perchlorate relates to the change in our awareness of its presence and persistence in the environment, its potentially severe health risks at low concentrations and the possibility of its natural formation. To be more precise, what makes perchlorate emerging from a forensics standpoint is our recent awareness of the existence of different natural sources of a contaminant that was not long ago considered entirely anthropogenic.

Before we proceed any further, it should be noted that perchlorate(s) is a generic term used to refer to the perchlorate ion provided by perchloric acid ($HClO_4$) or any salt of perchloric acid. Most of the perchlorate salts are very soluble and dissociate in water, freeing the perchlorate ion.

1.3   Keys to Forensic Investigations

Forensic studies, in general, involve reconstruction of past events based on the evidence that is left behind. In environmental forensics, the events refer to contamination release, and the evidence left could be the contaminant itself or any trace that the contaminant leaves behind. Additionally, any process, un-related to the release event, but resulting in the formation of the contaminant, should be recognized since it could interfere with the forensic evaluation. It becomes clear that understanding the contaminant and its environmental fate is the key to forensics reconstructions. This understanding involves: adequate methods and already existing knowledge. While the forensic investigative methods are well defined and evolving, the existing knowledge is limited for emerging contaminants. In such a situation, an increased awareness and an ability to predict changes are essential.

Subsequently, the keys for successful forensic investigations applicable to emerging contaminants such as perchlorate include:

- understanding the structure and physical-chemical characteristics of the contaminant;
- linking the contaminant characteristics to available environmental observations on the particular contaminant with the aim of identifying and predicting the environmental behavior and fate;
- based on its characteristics and available published information, understanding the uses and potential sources of the contaminant, and also its possible natural formation;
- linking the established information on the main sources/uses to the available information on environmental occurrences to identify new possible sources/formation mechanisms;
- using a variety of forensic methods in order to confirm or exclude the potential sources and determine release ages.

Above all, a successful forensic study should comprise both the application of available forensic investigative methods and the recognition of new possible scenarios. Any forensic investigation should begin with an open mind and the forensic conclusions should only be drawn after all possible scenarios are in-vestigated, along with any possible mechanism of contaminant formation. It is our human nature to blame the obvious. However, as in many detective stories and movies, it is not always the obvious that causes the problem. This is where sound science is needed, along with a comprehensive awareness and objective evaluation of many possible scenarios.

This chapter is structured to address (in each section) the main keys and steps of efficient forensic investigations.

2   Environmental Forensic Investigation of Perchlorate

2.1   Perchlorate – Unique Chemical

Perchlorate is a unique chemical. This is because of a combination of exceptional properties that explain its many uses in chemical analysis and many industries. Any forensic investigation should start with a comprehensive understanding of physical-chemical properties of a compound, based on which environmental behavior can be predicted and accurately interpreted. In this part of the chapter, we are providing the general characteristics of perchlorate and emphasize those characteristics that influence the environmental behavior of perchlorate and may have an impact on forensic evaluations (Table 1).

Perchloric acid is one of the strongest acids known. Most of its metal salts are soluble in water. Perchloric acid has low and insignificant volatility and little chemical reactivity under normal conditions. To understand the characteristics of this unique chemical, let us start with the structure of perchlorate. Perchlorate contains one chlorine atom bound to four oxygen atoms. The chlorine atom is in the most oxidized form, hence it cannot be oxidized any more. However, at least in theory, perchlorate can be reduced (reaction thermodynamically stable) to form chlorine atoms, yet the process is not observed to occur under normal environmental conditions. Due to such a lack of chemical reactivity, perchlorate is fairly stable (inert) and could persist in the environment at low concentrations for decades or more.

Perchloric acid is more effective as a decomposition reagent than either sulfuric or hydrochloric acid.[4] Perchlorate is a strong oxidizing agent which reacts violently with organic matter. This is why anhydrous perchloric acid is highly unstable against explosion and is also exceedingly corrosive. However, aqueous perchloric acid solutions are stable to heat and shock in the absence of any reactive material (organic compounds). Most perchlorate salts also present explosion risks. In general, the organic salts of perchlorate are under the highest risk of explosion, since they contain both the igniter (perchlorate ion) and the fuel (organic part). In general, the more covalent bonds are present in the structure of organic perchlorate salts, the higher is the risk of explosion. Certain heavy metal perchlorates (inorganic perchlorates) are also reported to have a high risk of explosion (if in contact with organic solvents). Aqueous perchlorate solutions containing heavy metal ions or organic matter should not be evaporated to dryness by heating. Also, any fat, oil or volatile organic substance (not miscible with perchloric acid) should first be decomposed with sulfuric and nitric acids before reaction with perchloric acid can occur safely.[4]

Wet oxidation with perchloric acid mixtures started to be employed in the 1930s. Perchlorate also has many uses in chemical analysis: as a digestion

Table 1 Physical-chemical characteristics of perchloric acid and perchlorates.

Characteristic	*Environmental effect*
Low vapor pressure → low volatility (not a volatile compound) **Little adsorption to soil:** ClO_4^- is negatively charged and will be repulsed by negatively charged soil particles	If perchlorate is released at or close to the surface, very little perchlorate will vaporize, hence most of the released perchlorate will percolate through soil and other porous media with minimal adsorption and will reach the groundwater
Highly soluble in water (except $KClO_4$)	• Dissolution is expected when in contact with water; perchlorate salts tend to dissociate completely in water and travel with groundwater as dissolved species long distances in space and time • Perchlorate salts will dissociate in water and perchlorate ion will predominate in solution
Density higher than water (around 1.91)	• Perchlorate will sink in water; it has the propensity to form DNAPL plumes (dense non-aqueous phase liquid plumes) that tend to travel vertically to lower depths
Exceptional acidic strength (stronger than most strong acids such as HCl and H_2SO_4)	Will dissolve and leach out metals and other inorganic and organic compound from the encountered environment
Powerful oxidizing agent (especially at high concentration and with heat) – reacts violently with organic matter (if concentration is appropriate):	Risk of explosion when in contact with organic substances or vapors (perchlorates are not stable if highly pure); heat will increase this risk Highly explosive perchlorates include:
• Use as explosive – advantages as compared with TNT or other nitroglycerine compounds, including: less affected by freezing, safer to handle, less sensitive to shock • Use in wet oxidation and digestion procedures for chemical analysis	• Organic perchlorates (diazonium or hydrazine perchlorate, perchlorate esters of aliphatic alcohols) • Fluorine perchlorate • Silver perchlorate ($AgClO_4$)
Low chemical reactivity (despite the theory that suggests reduction): • Not oxidizable • Not readily reducible in dilute solutions (high activation energy) • Weak metal ion complexing ability	Persist in environment (especially under aerobic conditions) for decades or more; depending on the groundwater velocity, perchlorate detected in certain wells could have been introduced and traveled with groundwater for 50 years or more
Solubilizing agent for a large range of organic contaminants (*e.g.* membrane-bound proteins, cellulose, acrylonitrile polymers, alcohols, ketones, amines)	While migrating through soil, it may dissolve a series of organic contaminants encountered on the way

Table 1 (*Continued*).

Characteristic	Environmental effect
Pronounced ability to form stable ion complexes with large symmetrical cations	May be bound stably to different compounds
Affinity for moisture without loss of porosity or ease of handling (does not become sticky); it **absorbs:**	Released solid perchlorate salts will absorb moisture
• Water • Ammonia • Alcohols and other polar compounds	
Chemical reactions of perchlorates may generate fire and explosion hazards	Perchloric acid reacts with alcohols and other organic compounds to form very unstable perchlorate esters, hence perchloric acid spills should be diluted immediately with water

medium for both organic and inorganic materials, an extraction medium for nucleic acids and a great variety of substances, precipitating and desiccating agent (more details are given in the next sub-section).

Table 1 highlights the main characteristics of perchlorate compounds and links them to the observed environmental fate and transport. Most of the facts included in the 'Environmental effect' column have been well established in the literature, confirming the good correlation between structure, characteristics and environmental behavior for perchlorate. Here, we do not give references to the particular studies confirming various environmental observed effects of released perchlorates since such evidence is overwhelming and the specific information will deviate from our main purpose of providing a forensic guide.

Due to the main characteristics described above, perchlorate is refractory to degradation and remediation in general. Two main approaches are mostly used for perchlorate remediation: (1) ion exchange and (2) bioremediation (through reduction in anaerobic conditions).

2.2 Sources of Perchlorate

The structure and main characteristics of perchlorate established and detailed in the previous sub-section enable us to understand and predict the environmental fate of perchlorate compounds released in the environment. This represents the first step in a forensic investigation. The next step relies also on a good understanding of the structure and characteristics of perchlorate compounds and represents the identification of perchlorate potential sources.

Site-specific data should be used through a detailed review of available historical publications. Additionally, a general understanding of the sources and associated uses of perchlorate is useful in identifying the particular sources that may be present in the particular case study (resulting through document review). For example, the main up-to-date confirmed sources presented in Table 2 could be checked (one by one) as potential contributing sources in a certain case. Information related to the actual existence of such sources should be identified through a first review of available site data that will confirm or rule out the many theoretical possible sources. The detailed review of historical data can then be targeted to focus only to the identified confirmed (based on site-specific information consulted) sources. Thus the whole review process becomes more efficient and likely to succeed simply by having a comprehensive list of potential up-to-date sources.

Table 2 provides a comprehensive and up-to-date list of both confirmed and suspected perchlorate sources and specific uses organized as a forensic reference guide. To our knowledge, there is limited literature information related to presenting and organizing the various sources of perchlorate based on the relevance for forensic study. Again, it is not within our scope to provide detailed references to support the information presented in Table 2. References will be provided in the next section related to the main forensic methods and published studies relevant for the forensic identification of perchlorates. Although not targeting forensic purposes, a comprehensive listing of perchlorate sources and uses has also been recently done by the Interstate Technology Regulatory Council (ITRC).[5] This list is recommended to be consulted to obtain more details on particular uses of perchlorate.

As seen in Table 2 and in accordance with up-to-date literature cited in the table,[1–18] perchlorate has a large variety of anthropogenic sources. At the same time, our awareness of the existence of natural sources is increasing with the detection of perchlorate in remote areas. Basically, we begin to understand that all that Nature needs in order to produce perchlorate is similar to what we need, namely a chlorine or oxychlorine precursor in solution and an energy source. The source of energy could be generated by lightning, UV radiation or heat, while the chlorine and oxychlorine precursors are abundantly generated in nature through chlorine evaporation from oceans and seas. If formation is to occur in the underground environment, a metal (or silica particle) is also important. Once these conditions required for natural formation of perchlorate are identified, it becomes clear that they could all be easily and abundantly generated in Nature (see also the discussion in the next section). Thus, natural perchlorate formation is not a matter of possibility, but just of appropriate conditions.

With so many potential sources, both anthropogenic and natural, it is no wonder why perchlorate is raising emerging issues and why it is perhaps one of the most challenging compounds from the forensic perspective. Every site should be evaluated for potential natural formation of perchlorate, keeping in mind the general conditions that facilitate perchlorate formation. We hope that this chapter will help to provide the knowledge for identification of naturally formed perchlorate.

Table 2 Guide for perchlorate source evaluation.[a]
(a) Anthropogenic sources

Source	Role/use of perchlorates	Associated perchlorates (examples)

Military activities Solid rocket fuel Perchlorate = propellant that is mixed with fuels	Propellant/oxidizer	• Solid propellants – 2 types:
		   • Composite (oxidant particles in a fuel matrix)
		   • Homogeneous (colloidal mixtures of oxidant and fuel)
		 • Liquid propellant mixture=a suspension in nitromethane

Fuels may include polysulfide rubber, hydrocarbon rubbers, epoxy resins and polyester resins		NH_4ClO_4 – commonly used in solid and liquid propellants $KClO_4$ – in composite propellants (higher burning rate) Various organic perchlorates
Munitions/explosives	Propellant/oxidizer	• Usually a mixture of either NH_4ClO_4 (most powerful blasting effects) or $KClO_4$ with sulfur and/or various organic materials • $NaClO_4$ – may also be used (*e.g.* photoflash) • Organic perchlorates – contain both the oxidant and combustible material (*e.g.* perchlorates of guanidine, dicyanodiamidine, aniline, pyridine, methylamine, hydrazine)
Pyrotechnics/signal flares (slow burning compositions)	Propellant/oxidizer	• $KClO_4$, NH_4ClO_4 – in signal flares • $KClO_4$ and $Ti(ClO4)_2$ – in firecrackers • $Zr(ClO4)_2$ – in pyrotechnic lacquer for primers • $Sb(ClO4)_2$ – in time delay pyrotechnic compositions

- Organic perchlorates
- Mixtures of ammonium perchlorate and sulfamic acid – to produce a dense smog or fog

Industrial activities

Activity	Function	
Fireworks	Oxidizer	$KClO_4$ ($SrCl_2$, $BaCl_2$ and $CuCl_2$ also present)
Highway safety flares	Oxidizer	$KClO_4$, $NaClO_4$, $Ba(ClO_4)_2$
Blasting agents – explosives in construction, mining	Oxidizer	$KClO_4$, NH_4NO_3
Match manufacturing	Oxidizer	$KClO_4$ blend with S as fuel
Smoke-producing compounds	Oxidizer	$KClO_4$, NH_4NO_3
Batteries and voltaic cells	Electrolyte	$LiClO_4$, $Mg(ClO_4)_2$, $Al(ClO_4)_3$, $Zn(ClO_4)_2$
Automobile air bags/ejection seats/aircraft fire extinguisher/aircraft oxygen	Gas (O_2) generator/initiator	$KClO_4$, $LiClO_4$, $NaClO_4$ or NH_4ClO_4 + strontium azide + boron
In charcoal briquettes	Igniter	$KClO_4$ + metal nitrates + carbonized matter
Hypochlorite/bleach solutions	Impurity	Massachusetts Department of Environmental Protection identifies this source as having a potential significant impact (one of the three main sources accounting for the many perchlorate occurrences in MA)[6]
	Potential formation through photo-oxidation in aqueous solutions	$Fe(ClO_4)_3$ and $KClO_4$ mixtures $NaClO_4$
Thermal reservoir pellets	Heat-generating material	Different perchlorates
Lubricating oils	Additive	
Metallurgy	Part of brazing and welding flux	
Paper and pulp processing	Bleaching powder	
Pharmaceutical industry	Compounding and dispensing agent	
Perchlorate, chlorate and chlorite manufacturing and disposal	Impurity	
Electroplating/electropolishing operations	Electrolyte	$HClO_4$ with acetic anhydride $HClO_4$ with ethanol or methanol $NaClO_4$ and acetic acid Other metal perchlorates, *e.g.* $Pb(ClO_4)_2$

Table 2 (*Continued*).

Source	Role/use of perchlorates	Associated perchlorates (examples)
Carbon fuel cell	Electrolyte	Concentrated $HClO_4$, $LiClO_4$
In solutions preventing corrosion of iron and steel surfaces		$HClO_4$
Oil and gas production	Brine water intrusion	
Well drilling	Part of permeability aid agent	
Paint and enamel manufacturing	Curing/drying agent	
Ammonia production	Catalytic mixture	
Poly(vinyl chloride) (PVC) manufacturing	Catalyst	
Photography/photographic emulsions	Sensitizing agent	Quaternary ammonium perchlorates
Production of semiconductor devices	Component of etching solutions	$HClO_4$
Etching steels and stainless steels	Part of etchants	Aqueous perchloric solution and compounds of Ce(IV)
Metamphetamine laboratories		Perchlorates are associated with red phosphorus
Textile industry	Fixer for fabrics and dyes; textile bleaching agent	
Chemical industry (more details in literature1,4)	Chemical analysis: • Dissolution and oxidation of inorganic samples (including metals and alloys) • Titrimetric reagent • Precipitation and extraction • Deproteinization agent • Drying agent, also remove small amounts of polar compounds from inert gases	$HClO_4$ (in combination with other acids such as HNO_3 or H_2SO_4) – as digestion agent in wet chemistry $HClO_4$ – quantitative extraction of nucleic acids $HClO_4$ (precipitation agent for potassium) $NaClO_4$ (precipitation of alkaloids) $Mg(ClO_4)_2$, $Ba(ClO_4)_2$ – drying agents
	Solvent	For inorganic and organic materials (alcohols, ketones, amines, resins)

Source	Mechanisms of formation	Average detected/references
	Organic synthesis, esterification, acetylation and polymerization; also isomerization	$HClO_4$ – highly effective catalyst in acetylation of cellulose
		$HClO_4$, acetyl perchlorate and metal perchlorates – catalysts in polymerization
Medicine (past uses)		
Medicine against hyperthyroidism	Thyroid inhibitor	
Dentistry	Drilling agent and disinfectant in the treatment of dental canals	$HClO_4$
Drugs and animal food supplements		
Thyreostatic drugs – thiouracil type	Thyroid inhibitor	
Animal feed additives – research by Russian scientists shows that perchlorate is metabolized and completely eliminated from animals in 24–46 h	Weight stimulants due to thyrostatic effect	NH_4ClO_4, $NaClO_4$, $KClO_4$
Agricultural activities		
Applications of Chilean nitrate	Impurity in fertilizer	$KClO_4$ (1.7–7.7 g kg^{-1})
Chlorate defoliants	Impurity	
	Photo-oxidation product?	$NaClO_3$ – used in herbicides

(b) Natural sources

Source	Mechanisms of formation	Average detected/references
Natural mineral deposits		
Nitrate deposit from Atacama Desert, Chile – the most studied natural source	Atmospheric formation and deposition in arid climates:	Up to 6.79%
	Reaction: Cl$^-$ (sea salt) + O_3, followed by: transport and evaporative concentration in unsaturated zone (correlation with iodate)	Average 1–1.5% in salpeter[1] 0.03–0.1% along with 6.3% nitrate in deposits[2]
Canadian and New Mexico potash ore (KCl) from sylvinite deposits	Probable atmospheric formation and deposition (in the absence of soil leaching in arid regions)	25–3741 mg kg^{-1} (Refs. 3,7)
Californian hanksite from evaporative deposits in Sears Lake, CA		280–285 mg kg^{-1} (Ref. 3)

Table 2 (*Continued*).

Source	Mechanism of formation	Average detected/references
Bolivian Playa Crust Caliche formation from Death Valley, CA		489–1745 mg kg^{-1} (Ref. 3) Confirmation in 1988 study[8]
Environmental media		
Rainwater, snow, seawater, atmospheric aerosols, groundwater, bottled water	Possible atmospheric formation Sonication of seawater	10–200 ng l^{-1} in rainwater[9,10]
Kelp		885 mg kg^{-1} (Ref. 3)
Oxychlorine precursors		
Oxychlorine precursors in aqueous solution (hypochlorite, chlorite and chlorate)	Photochemical formation from aqueous oxychlorine anions – by exposure to UV light (or sunlight) of hypochlorite, chlorite and chlorate solutions	A recent study[11] reported formation of: • 5, 25 and 626 µg l^{-1} perchlorate from 100, 1000 and 10000 mg l^{-1} of chlorite solution, respectively • 1491 µg l^{-1} from 10000 mg l^{-1} hypochlorite solution • 468 and 1522 µg l^{-1} from 1000 and 10000 mg l^{-1} chlorate solution, respectively

Subsurface environments with buried metallic objects

Source	Mechanism of formation	Average detected/references
Chlorinated water from steel tanks (UST) with corrosion protective systems	Electrochemical generation in subsurface Generally, electrochemical formation in subsurface from thunderstorms on metallic buried objects (cast iron) involves: Cl$^-$ → HOCl/ClO$^-$ → ClO$_3^-$ → ClO$_4^-$ Specifically (proven situation), electrolytic generation from aqueous solutions of precursor ions (*i.e.* Cl$^-$, ClO$^-$, ClO$_2^-$, ClO$_3^-$) under pH: 6–8, with an applied voltage and in the presence of cast iron, titanium or niobium anode	74 µg l^{-1} perchlorate in water stored in a steel tank in Texas[13,14]

(c) Unconfirmed sources/formation

Source	Role/mechanisms of formation	Examples/references
Anthropogenic		
Shotgun firing activities	• As propellant – mixture of Al and K perchlorate (although not commercially available, there is evidence that shooters may use such combination for a stronger/powerful gun load) • Also as primers – some typical igniters (formulations such as Red Tracer, R-257)	Encyclopedia of Explosives and Related Items[15] • http://enviro.navy.mil/erb/erb_a/restoration/technologies/remed/uxo_remed/rpt-sar-imple.pdf • http://en.wikipedia.org/wiki/Percussion_cap • http://www.totse.com/en/bad_ideas/guns_and_weapons/reloads.html • http://www.logicsouth.com/~lcoble/stuff/shotload.txt
Wide range of common garden fertilizers (others than Chilean nitrate)	Some production treatment common to all fertilizers, such as oils used to keep the fertilizer dry or brine solution used to control acidity, could be the source of *perchlorate contamination*; data are so far inconclusive	Ubiquitous perchlorate reported in a large range of fertilizers at concentrations from 0.15 to 0.84 wt.% • http://pubs.acs.org/hotartcl/est/99/oct/oct-news5.html See also a 1999 study[16]
Municipal landfill	Leachates with perchlorate may be generated due to the increasing use of perchlorate salts in common household and commercial products	MA DEP pointed out the need to investigate such occurrences[6] (2006)
Leather tanning	Tanning and finishing agent	
Aluminum refining, aluminum electropolishing		Unconfirmed by recent research
Rubber manufacturing	Rubber as a binder in motor rockets	
Nuclear reactor	Nuclear warheads on rockets	

Table 2 (*Continued*).

Source	Role/mechanisms of formation	Examples/references
Nuclear reactors, electronic tubes, transformers	Dielectric	
Natural		
Natural nitrate deposits used for fertilizers other than Chilean nitrate	Possible atmospheric formation followed by deposition	Contradictory studies[17]
Natural minerals, apart from those with already confirmed detection		Mission Valley Formation (San Diego, CA) – carbonate-rich layer[18]
Swimming pools – natural formation in swimming pools treated with hypochlorite products	Through photochemical oxidation of hypochlorite	Identified as a possibility in need for research by MA DEP[6]
Atmospheric sources/processes	Electrical discharge (lightning) through NaCl aerosol (precursors: Cl^- and ClO^-) Photochemical oxidation of aerosol perchlorate precursors (see discussion before[11])	Mechanism of formation defined though recent study[12]
Oxychlorine precursors in aqueous solutions	Direct chemical oxidation of chlorate by strong oxidizer Reaction of a strong mineral acid on a chlorate Thermal decomposition of chlorate	

[a]Blank spaces: information not available.

2.3 Tracking Perchlorate in the Environment

A large variety of forensic techniques are currently available to investigate the source and age of perchlorate in the environment. A list of these techniques along with guidance for their use in perchlorate forensic investigations is provided in Table 3. Recent successful applications of such techniques are referenced (where available) in the same table. Details related to many of these forensic techniques are provided in several excellent forensic text books,[19–22] as well as in different issues (2000-present) of the quarterly peer-reviewed Environmental Forensics Journal available on-line (http://www.informaworld.com/smpp/title~db= all~content=t713770863). An excellent review article[22] on perchlorate forensics is also available in the text book of Morrison and Murphy. However this is one of the few review articles to date specifically addressing perchlorate forensics.

The information presented in Table 3 relates to the general overview of forensic investigative methods applicable to perchlorate and useful forensic hints related to each of these methods. It is recommended to use more than one forensic method in order to obtain so-called independent lines of evidence. Basically, a strong forensic case is built only when the conclusions of several independent forensic methods concur. This is why we recommend the use of as many techniques as possible to solve any forensic issue, regardless of how obvious the final conclusions may appear.

Any forensic investigation (regardless of the type of contaminant involved) should start with a thorough document review. This may provide the necessary information to solve the forensic 'mystery' in many cases. If monitoring data are available, forensic fingerprinting techniques (*e.g.* diagnostic ratios), hydrogeological modeling and statistics could and should be employed to study and interpret existing monitoring data. Often, getting new samples is not an option, for several reasons (contaminants are already gone, inaccessibility to the site or difficulty in obtaining approvals for sampling). However, even when new sampling is possible, it is recommended first to use the site's existing data. If this may not help in reaching forensic conclusions by itself, it is definitely useful in designing the forensic sampling plan and strategy for analyzing new data.

A distinct column in Table 3 is dedicated to criteria for evaluating the natural formation of perchlorate. This is because of the possibility (at least in theory) of having natural perchlorate formation and contributions in basically any environment. The first indication that perchlorate can be formed independent of human activity was related to the discovery of significant amounts of perchlorate associated with the nitrate deposits in the Atacama Desert in Chile. Such an occurrence was confirmed through many studies[1,2] showing that the fertilizers from those Chilean nitrate deposits, shipped all over the world, are confirmed sources of perchlorate. This observation had significant forensic implications, since Chilean deposits provide a source for testing natural perchlorate. Thus, once the perchlorate from Chile nitrate deposits was confirmed as natural, it started being tested and compared with anthropogenic perchlorate. The aim of such studies was to find distinct traits between

Table 3 Forensic techniques applicable to perchlorate.

Forensic method	Applications	What to look for	Criteria to evaluate natural formation	Application examples/references
Historical document review (first step of any forensic investigation)	• Infirm/confirm potential sources • Establish site-specific background values → identify impacts above background • Evaluate any available monitoring data – sometimes sufficient to answer forensic questions	• Information available related to any of the potential sources as defined in Table 2 → establish site-specific potential sources • Site monitoring data comprising physical-chemical and geological analytical information available (check governmental agencies files such as Water Board or DTSC) • Published background values (for contaminants of concern) – e.g. consult USGS, EPA, DTSC and other governmental agency files • Contaminants used at the site through consulting agency's files such as fire department or city/municipality files • Reported and suspected spills/accidents or incidents that could have resulted in contaminant released at the site and surrounding sites • Historical aerial photographs showing changes in land use, structures, etc.	• Geographical settings of the site (e.g. is it in an arid area close to the ocean?) • Occurrence correlation with other ions and with potential source area • Historical existence of underground cathodic protection systems and associated USTs • Occurrence ranges vs. published natural occurrences in other areas • The lack of any recorded human related activities and/or releases in the area (see studies related to perchlorate occurrence in West Texas[24-25,27])	General application described in Murphy and Morrison[19,20]
Isotopic fingerprinting	• Distinguish atmospherically formed perchlorate • Distinguish between sources of perchlorate (e.g. anthropogenic vs. natural or within each category)	• Deviation in oxygen isotopic composition ($\Delta^{17}O$) • Isotopic composition for oxygen (^{18}O, ^{17}O), chlorine (^{37}Cl) and hydrogen (^{2}H) stable isotopes of studied samples and comparison samples (from potential sources) • Stable isotopic composition for strontium isotopes • Plotting the isotopic values against each others • Evaluate biodegradation occurrence and predict its effects on isotopic fractionation	• Anomalous value of ^{17}O composition clearly indicate natural atmospheric formation such as in the case of Chilean fertilizer • However, the lack of ^{17}O anomaly may not suggest non-natural perchlorate (could still be formed in atmosphere but not involving ozone) • Comparison with isotopic composition of reference samples (with known naturally formed perchlorate)	• Unique oxygen isotopic signature for atmospherically formed perchlorate – involving ozone oxidation[23] • Isotopic fingerprinting of a variety of perchlorate compounds[25,28,30,31] • Isotopic fractionation during microbial reduction[29,30] • Use of Sr isotopes to distinguish between perchlorate sources[29,32] • NASA study related to forensic investigation of JPL's perchlorate[31]

Chemical fingerprinting: • Associate ions • Signature chemicals	Source identification by association with other ions and potentially signature compounds, and also by evaluating spatial distribution	• Spatial distribution with aquifer, well type and land use • Vertical profile of perchlorate and other ions • Perchlorate concentrations in saturated *vs.* unsaturated zone • Relation between perchlorate, depth to water and saturated thickness • Concentration of other ions from the sample • Identify and check for possible signature chemicals (associated with suspected sources) • Note: associate ions may be used to track perchlorate sources only for recent releases, since such ions (K^+, Na^+, *etc.*) have different fate and transport than perchlorate and do not persist in water	• Correlations with other ions (iodate, nitrate, sulfate, *etc.*) may indicate atmospheric formation • Correlation with nitrate may indicate natural formation using storm lighting to generate energy • The ratios Cl^-/Br^- and Cl^-/ClO_4^- may indicate atmospheric origin • Vertical profile of perchlorate is consistent with surface sources, including atmospheric formation and deposition • Possible correlation with TDS, anions and cations could indicate general flushing of salts from the land surface to water table by precipitation–irrigation–evaporation cycles	Evaluation of perchlorate occurrence in groundwater and unsaturated zone in West Texas and New Mexico[24,25,27]
Mineralogical fingerprinting	Identify and prove association with mineral formations	• Compare mineralogical composition with already proven natural formation identified as perchlorate-containing minerals • The general site conditions and any correlations between the sample composition and surrounding environments • The presence of similar mineralogical material in the broad study area	• Match with mineralogical fingerprints of minerals already known to contain perchlorate • Match conditions known to favor natural formation of perchlorate with mineralogical information on the samples	Discovery of a new perchlorate-bearing mineral in a carbonate-rich marine layer in southern California[18,22]
Evaluate *in situ* natural formation scenarios	Identify possible formation from oxychlorine precursors under appropriate environmental conditions (metal, current source, pH)	• Site characteristics • Presence of buried metallic objects such as UST with corrosive protection systems • The availability of current source (such as lightning strikes) • The presence and concentration of any perchlorate precursor (aqueous solution) • The pH and other site general parameters	• Any match that will facilitate conditions similar with those published for confirmed natural formation (see also Table 2) • Confirmation through laboratory studies if match is not found, but theoretical conditions for perchlorate formation are identified	Evaluation and laboratory demonstration study of photochemical formation of perchlorate from aqueous oxychlorine anions[11] Review of mechanisms that could be involved in natural perchlorate formation in the atmosphere[12] Electrochemical generation of perchlorate in sub-surface

Table 3 (*Continued*).

Forensic method	Applications	What to look for	Criteria to evaluate natural formation	Application examples/references
				environments in the presence of buried metals and using thunder-storms energy was evaluated and demonstrated[13,14]
Groundwater modeling	• Determine upgradient areas and timing for perchlorate to reach the sampling point from different source areas	• Groundwater flow directions and velocities in the study area – timing for contamination from potential sources to reach the sampling area	• Direct evaluation cannot be done through this method alone	General application described in Murphy and Morrison[19,20] NASA study[31]
Statistical evaluation	• Identify and delineate perchlorate sources	• Chemical composition of samples (perchlorate and other compounds and elements from each sample) • Similarity of environmental conditions to which different samples have been exposed	• Direct evaluation cannot be done through this method alone	General application described in Murphy and Morrison[19,20]
Mass balance calculations	• Confirm/infirm occurrences due to agricultural activities	• Official documents related to the amount of fertilizers (from Chile) used over a certain area and within a certain time frame • Any study related to the composition of fertilizer used in the area → identify possible signature chemicals to be used in mass balance calculations also	• Maximum total amount that would have resulted from fertilizer application could be used to confirm or exclude this natural source	Evaluate perchlorate occurrence in groundwater in West Texas and New Mexico[24,25]

| Dendroecology (using tree ring data to age date and characterize environmental releases) This method has not yet been applied in perchlorate studies, but has high forensic potential | • Age date perchlorate releases with precision to the year, even when contamination events happened a long time ago (100 years or more), provided that trees of appropriate age are present
• Characterize the type of release: sudden and accidental vs. gradual leak
• Identify multiple releases | • Ring width anomalies
• Cl concentration peaks along the tree core sample
• Associated elemental peaks (Na, K, etc.) – to evaluate other possible contributions resulting in Cl elemental peak in the study tree
• Statistical data on ring width for similar species in the study area (for comparison)
• Climatic data for the study area – to evaluate potential climatic impact on tree growth | • In conjunction with site-specific information and isotopic data, this method can help delineate natural occurrences of perchlorate
• Indirectly, the method may help distinguish natural perchlorate occurrences by linking contamination to potential sources (trees located in between potential sources and sampling site) | General information on the technique and successful applications are presented in recent publications by Balouet et al.[33,34] |

anthropogenic and natural perchlorate. Such a trait proved to be the oxygen isotopic composition. Moreover, the distinct isotopic composition of natural perchlorate confirmed its atmospheric origin. Bao and Gu[23] have undoubtedly shown that perchlorate from Chilean nitrate deposits was formed in the atmosphere, involving ozone oxidation of chlorine evaporated from oceans, followed by deposition and evaporative concentration in arid environments. Moreover, this study discusses several mechanisms of natural perchlorate formation, proving that only one oxygen from ozone remains incorporated in the naturally synthesized perchlorate. This study demonstrated that perchlorate naturally formed in the atmosphere has a distinct isotopic composition consisting of the ^{17}O isotope anomaly. Basically, the atmospherically formed perchlorate retains (at least partially) the isotopic signature of ozone. The forensic implications of such distinct isotopic compositions are valuable. Simply by determining the oxygen isotopic composition (both for ^{17}O and ^{18}O isotopes) and calculating the deviation from the meteoric line, it is easy to determine if atmospherically formed perchlorate contributed (completely or in part) to the perchlorate from a certain environmental sample. However, it should be noted that perchlorate may be formed naturally in the atmosphere without involving ozone oxidation (as shown in Table 2). In such a situation, the anomaly in ^{17}O isotopic composition may not be noticeable. This may be the case with perchlorate detected in West Texas that is thought to be of natural origin, although its isotopic composition does not suggest any ozone involvement.[24-27]

Our discussion so far illustrates how one forensic method (isotopic analysis) may help distinguish between perchlorate sources (especially natural *vs.* anthropogenic). Another method applicable to track natural perchlorate from fertilizers is mass balance calculation. Thus, considering that the general amount of perchlorate in fertilizers based on Chilean nitrate is generally known and the use of such fertilizers could be tracked down, it remains a simple mathematical calculation problem to evaluate the total mass of perchlorate that could have been introduced into the environment through agricultural fertilization and thus exclude or confirm fertilizers as sources for a particular case study. Such an independent forensic line of evidence was used by Jackson's research group investigating the unusual occurrence of perchlorate in groundwater and soil of many counties in Texas.[24,25,27]

Following the discovery of perchlorate in Chilean nitrate fertilizers, studies were conducted to evaluate the possibility of the occurrence of perchlorate in other types of fertilizers. While the results generally show that such correlations (with fertilizers other than Chilean nitrates) could not be established, no systematic studies are known to have been undertaken in this respect. In the absence of such systematic studies, the possibility remains that perchlorate may be associated with other nitrate deposits, especially formed in similar geographical conditions to the Chile Atacama Desert.

Thus, apart from the perchlorate found in nitrate fertilizers from Chile, another important event raising our awareness of the existence of more natural perchlorate than what was initially thought is represented by the unusual detection of perchlorate over large areas (in soil and groundwater) in many counties (56) in the West Texas area, as mentioned before.[24,25,27] Jackson's group at Texas Tech University made a pioneering contribution in pointing out many mechanisms of natural formation of perchlorate. The problem is both intriguing and interesting, in that no anthropogenic contributions to the perchlorate detected in Texas could be identified as the main perchlorate source(s). Additionally, the isotopic signature of the Texas samples found that perchlorate was distinct from both anthropogenic and Atacama perchlorate samples.[26] This suggests that the perchlorate found in Texas could be:

- of natural origin, having a distinct isotopic signature from the perchlorate from nitrate fertilizers from Chile (may be formed through various other mechanisms including atmospheric formation without involving ozone – see Table 2);
- a mixture of synthetic with natural (from fertilizer) perchlorate in a ratio of 19:1.

To date, various mechanisms of natural formation of perchlorate have been investigated and some have been confirmed. For example, Jackson's group pointed out that perchlorate can be naturally formed in aqueous solutions of oxychlorine anions (*e.g.* hypochlorite, chlorite and chlorate) exposed to UV light.[11] However, the precursor solutions need to be of extremely high concentration (100–10 000 mg l^{-1}), which is usually not expected in natural environments. Additionally, the study showed that chloride solution was the only precursor that did not form perchlorate while exposed to light during the 7 day experimental period. Another interesting observation is that presumably potential perchlorate formation may occur in the absence of UV light (dark) via dark disproportionation reactions. The paper concluded that perchlorate could be a proposed sink and a significant reservoir (50%) of total inorganic chlorine in the lower stratosphere. An excellent article discussing different mechanisms of atmospheric formation of perchlorate has also been published by the same group.[12]

Apart from atmospheric formation, Jackson's group has identified and demonstrated another natural formation pathway for perchlorate in underground environments that contain buried metallic objects (such as underground storage tanks).[13,14] This formation involves electrochemical generation with the required energy provided by lightning.

From the discussed case studies, it can be stated that perhaps isotopic fingerprinting is the most efficient and best defined method used to date for the identification of perchlorate sources. Apart from the study of Bao and Gu,[23] related to the unique oxygen isotopic signature of perchlorate formed by ozone oxidation, Sturchio's group at the University of Illinois at Chicago have carried out many studies and have described the method and pointed out the significant differences in isotopic composition of natural (Chilean fertilizers) *vs.* anthropogenic perchlorate, and also of various anthropogenic sources.[23,26,28] At the same time, the isotopic composition of perchlorate provides an excellent insight related to the occurrence of biodegradation, which was proven to result in a significant isotope fractionation with an increase in the heavier isotopes.[29,30] An excellent application of isotopic fingerprinting for perchlorate source delineation was recently performed by NASA and used as an independent line of evidence while tracking the perchlorate from a city well downgradient of the Jet Propulsion Laboratory (JPL) facility.[31]

Apart from using the isotopic analysis of perchlorate component elements (such as chlorine and oxygen), useful forensic information can be acquired from strontium isotopes. Stable isotopes of strontium (^{87}Sr/^{86}Sr) have been used to trace groundwater migration and identify contaminant sources, by quantifying the effects of seawater intrusion and oilfield brines. Strontium is used as a surrogate chemical to identify a variety of contaminants of concern in water. The method has high precision (0.002%) and results to date indicate significant differences among various perchlorate sources.[32]

Apart from isotopic fingerprinting, another forensic method holds great potential for perchlorate source delineation and age dating. This is an emerging forensic application with high age dating precision (to the year), being independent of the physical presence of contamination at the moment of sampling. The method is called dendroecology or the use of tree rings to age date and characterize environmental releases. Balouet and Petrisor (the latter of Environmental International, Paris) have been involved in successfully applying the dendroecology method to age date and characterize a variety of contaminants. More information on the method and its application can be found in recent publications.[33,34] The method is based on well-established dendrochronology principles that link the width of yearly tree growth rings with climate and any environmental factors affecting the growth. When a contaminant enters the tree, it may affect the growth, which becomes reflected in the width of the ring (becoming narrower). Moreover, such growth anomalies could be linked with the concentration of different elements, more precisely with elemental peaks in certain rings, in order to confirm contamination events. For example, compounds containing chlorine (such as chlorinated hydrocarbons) will determine the appearance of a chlorine peak corresponding to the tree growth ring of the year when the contaminant entered the tree (through the root system). The method was successfully applied to different fossil fuels (*e.g.* diesel and fuel oil) and chlorinated solvents. Although the method has not yet been applied to track and age date perchlorate releases, it holds great potential in this respect. The method is of particular interest for environments with shallow groundwater (above 30 ft below ground surface) and at sites where old trees are present close (downgradient) to potential source(s). Apart from the high precision, the method has the advantages of being environmental friendly, easy to apply (simple sampling device) and less expensive compared with classic fingerprinting techniques for soil and water.

As discussed and shown in Table 3, a large variety of forensic methods are available for perchlorate forensic studies. The use of independent lines of evidence is the best forensic option providing reliable evidence.[31]

3 Conclusions

This chapter provides forensic guidance for the delineation of the sources and fate of perchlorate in the environment. Up-to-date information on general characteristics, sources and environmental tracking methods for perchlorate has been presented. Emphasis was given to the various mechanisms of natural formation of perchlorate. Such formation could have tremendous impacts in any forensic investigation and needs to be carefully evaluated before a forensic conclusion is drawn. Basically, if perchlorate precursors and an energy source (voltage, UV light, lightning, *etc.*) are present, natural formation of perchlorate may occur. It is the role of forensic scientists to understand and distinguish such occurrences from anthropogenic releases and background values. The information presented here provides the basis for perchlorate forensic investigations, since, as Louis Pasteur said, 'in the field of observations, chance favors the prepared mind'.

References

1. A. A. Schilt, *Perchloric Acid and Perchlorates*, 2nd edn., G. Frederick Smith Chemical Company, Powell, OH, 2003, p. 300.
2. E. Urbansky, S. Brown, N. Magnusom and C. Kelty, *Environ. Pollut.*, 2001, **112**, 299–302.
3. G. Orris, G. Harvey, D. Tsui, J. Eldridge, *Preliminary Analyses for Perchlorate in Selected Natural Materials and Their Derivative Products*, US Geological Survey, Open File Report 03-314, US Department of the Interior, Washington, DC, 2003, p. 6.
4. A. A. Schilt, *Perchloric Acid and Perchlorates*, G. Frederick Smith Chemical Company, Library of Congress, 1979, Catalog card number 79-63068.
5. ITRC, *Perchlorate: Overview of Issues, Status and Remedial Options*, ITRC, 2005, Washington, D.C.
6. Massachussetts Department of Environmental Protection, *The Occurrence and Sources of Perchlorate in Massachussetts*, Draft Report August 2005, updated April 2006.
7. P. Renne, W. Sharp, I. Montanex, T. Becker and R. Zierenberg, *Earth Planet. Sci. Lett.*, 2001, **193**, 539–547.
8. G. Ericksen, J. Hosterman and St. P. Amand, *Chem. Geol.*, 1988, **67**, 85–102.
9. D. M. Murphy and D. S. Thompson, *Geophys. Res. Lett.*, 2000, **27**, 3217–3220.
10. S. A. Snyder, B. J. Vanderford and D. J. Rexing, *Environ. Sci. Technol.*, 2005, **39**, 4586–4593.
11. N. Kang, T. A. Anderson and W. A. Jackson, *Anal. Chim. Acta.*, 2006, **227224**, 1–9.
12. P. K. Dasgupta, P. K. Martinelango, W. A. Jackson, K. Tian, R. W. Tock and S. Rajagopalan, *Environ. Sci. Technol.*, 2005, **39**, 1569–1575.

13. R. Tock, W. Jackson, R. Rainwater and T. Anderson, *Potential for Perchlorate Generation in Drinking Water Storage Tanks*, Texas Water Development Board, Austin, TX, 2003.
14. R. Tock, W. Jackson, T. Anderson and S. Aunagiri, *Corros., Corros. Eng. Sect.*, 2004, **60**, 757–763.
15. B. T. Fedoroff and G. D. Clift, *Encyclopedia of Explosives and Related Items*, Dover, NJ, Picatinny Arsenal, 1960.
16. S. Susarla, T. Collette, A. Garrison, N. Wolfe and S. McCutcheon, *Environ. Sci. Technol.*, 1999, **33**, 3469–3472.
17. E. Urbansky, T. W. Collette, W. P. Robarge, W. L. Hall, J. M. Skillen, P. F. Kane, *Survey of Fertilizers and Related Materials for Perchlorate (ClO_4^-)*. *Final Report*, 2001, Nat. Risk Management Research Laboratory, U.S. EPA, Cincinatti, OH.
18. B. Duncan, R. Morrison and E. Vavricka, *Environ. Forensics*, 2005, **6**, 1–11.
19. B. L. Murphy and R. D. Morrison (eds.), *Introduction to Environmental Forensics*, 2nd edn., Elsevier Academic Press, 2002.
20. B. L. Murphy and R. D. Morrison (eds.), *Introduction to Environmental Forensics*, 2nd edn., Elsevier Academic Press, 2007.
21. R. D. Morrison and B. L. Murphy (eds.), *Environmental Forensics. Contaminant Specific Guide*, Elsevier Academic Press, 2006.
22. R. D. Morrison, E. A. Vavricka, P. B. Duncan and Perchlorate, in *Environmental Forensics. Contaminant Specific Guide*, ed. R. D. Morrison and B. L. Murphy, Elsevier Academic Press, 2006, pp. 167–185.
23. H. Bao and B. Gu, *Environ. Sci. Technol.*, 2004, **38**, 5073–5077.
24. W. A. Jackson, K. A. Anandam, T. Anderson, T. Lehman, K. Rainwater, S. Rajagopalan, M. Ridley and R. Tock, *Groundwater Monit. Remed.*, 2005, **25**, 137–149.
25. S. Rajagopalan, T. A. Anderson, L. Fahlquist, K. A. Rainwater, M. Ridley, W. A. Jackson, *Environ. Sci. Technol.*, 2006, Published on the web 04/07/2006.
26. N. C. Sturchio, J. K. Bohlke, B. Gu, J. Horita, G.M. Brown, A. D. Beloso, L. J. Patterson, P. B. Hatzinger, W. A. Jackson and J. Batista, Stable isotopic composition of chlorine and oxygen in synthetic and natural perchlorate, in *Perchlorate: Environmental Occurrence, Interactions and Treatment*, ed. B. Gu and J. D. Coates, Springer, Berlin, 2007, Chapter 5, pp. 93–110.
27. B. Rao, T. A. Anderson, G. J. Orris, K. A. Rainwater, S. Rajagopana, R. M. Sandvig, B. R. Scanlon, D. A. Stonestrom, M. A. Walvoord and W. A. Jackson, *Environ. Sci. Technol.*, 2007.
28. J. K. Bohlke, N. C. Sturchio, B. Gu, J. Horita, G. M. Brown, W. A. Jackson, J. Batista and P. B. Hatzinger, *Anal. Chem.*, 2005, **77**, 7838–7842.
29. N. C. Sturchio, P. B. Hatzinger, M. D. Arkins, C. Suh and L. J. Heraty, *Environ. Sci. Technol.*, 2003, **37**, 3859–3863.
30. N. C. Sturchio, J. K. Bohlke, A. D. Beloso Jr., S. H. Streger, L. J. Heraty and P. B. Hatzinger, *Environ. Sci. Technol.*, 2007, **41**, 2796–2802.
31. T. K. G. Mohr, *Southwest Hydrol.*, July/August 2007.

32. R. Hurst, The role of naturally occurring isotopes in forensic investigations of perchlorate-impacted soil and groundwater. Presented at Perchlorate in California's Groundwater: 11th Symposium in GRA's Series on Groundwater Contaminants, Groundwater Resources Association, Glendale, CA, 4 August 2004.

33. J. C. Balouet, G. Oudijk, I. G. Petrisor and R. Morrison, in *Introduction to Environmental Forensics,* 2nd edn., ed. B.L. Murphy and R.D. Morrison, Elsevier, Amsterdam, 2007, pp. 699–731.

34. J. C. Balouet, G. Oudijk, K. Smith, I. G. Petrisor, H. Grudd and B. Stocklassa, *Environ. Forensics,* 2007, **8**, 1–17.

Tracking Chlorinated Solvents in the Environment

IOANA G. PETRISOR AND JAMES T. WELLS

The present is the key to the past.

James Hutton

1 Introduction – The Environmental Legacy

The abundance and persistence of chlorinated solvents in many natural environments brought this class of compounds into the spotlight of both scientific and legal communities. Chlorinated solvents are the most frequently detected groundwater contaminants in the USA.[1] According to a 1989 estimate by the US Environmental Protection Agency (EPA),[2] trichloroethylene (TCE), a common chlorinated solvent, was detected at mean concentrations of $27.3\,\mu g\,l^{-1}$ [exceeding the $5\,\mu g\,l^{-1}$ maximum contaminant level (MCL) drinking water standard established by the EPA] in 3% of surface water samples and 19% of groundwater samples analyzed worldwide. Chlorinated solvents have been reported in the food supply, human blood and vegetation, including trees, although their presence in fruits has not been evidenced so far (see, for example, the recent study of Chard *et al.*[3]). Such widespread occurrences are the result of the common industrial and commercial uses combined with an initial lack of awareness of the risks to human health and the environment posed by chlorinated solvents. Today, we have to deal with the environmental legacy of chlorinated solvents. The best way to do so is to take advantage of advances in science and apply investigative tools to help understand the nature and extent of contamination and to better remediate the releases.

The most common chlorinated solvents are: perchloroethylene (PCE), trichloroethylene (TCE), 1,1,1-trichloroethane (1,1,1-TCA), carbon tetrachloride (CT), chloroform (CF) and methylene chloride (MC). From a historical

Issues in Environmental Science and Technology, No. 26
Environmental Forensics
Edited by RE Hester and RM Harrison

perspective, CT is the chlorinated solvent that first came into widespread use by the beginning of the 20th century. CT was later replaced by PCE and totally phased out in the 1990s under the Montreal Protocol.[4] In addition to CT, 1,1,1-TCA, was also phased out as a result of the Montreal Protocol (due to ozone destruction). This phase-out triggered the need for substitutes for 1,1,1-TCA, which has been achieved largely with TCE and PCE, which, ironically, were previously replaced by 1,1,1-TCA in the 1970s following the Clean Air Act. PCE remains the solvent of choice for 85–90% of approximately 30 000 dry cleaners and launderers in the USA.[4] However, the PCE releases from dry cleaners are less probable today since the modern cleaning equipment recovers 95–99% of used PCE[4] and our awareness of and capability for early detection and treatment are increasing.

Chlorinated solvents may pose a considerable health risk. For example, TCE was classified (by the EPA) as a probable human carcinogen in 1985. Due in large part to its potential carcinogenicity, the MCL for TCE is very low (0.005 mg l^{-1} or 5 µg l^{-1}). The effects of TCE continue to be evaluated and after a series of reviews by the National Academy of Sciences and the EPA, there is ongoing debate as to whether even this low level of exposure affords adequate protection.

Apart from their potential risk to human health, what makes chlorinated solvent releases problematic is their environmental persistence. The first alarming signal relating to chlorinated solvents' potential to persist in the environment was an isolated publication in the late 1940s in the UK by Lyne and McLachlan,[5] which is believed to be the first published description of environmental impairment by chlorinated solvent releases to the environment.[6] Through the 1940s, other records regarding the awareness of environmental impacts from the use of chlorinated solvents are, in general, scarce and the prevailing state of knowledge by manufacturers, distributors and users of chlorinated solvents remains a historical uncertainty. A recent debate on this topic was held in the journal *Environmental Forensics*. According to a paper in that journal in 2001, contamination with chlorinated solvents in Southern California was foreseen by the 1940s,[7] while a response to that paper[8] argues that the potential impacts of chlorinated solvents was simply not widely appreciated in the 1940s. Beyond this debate, the fact remains that there are relatively few references in the literature regarding environmental impacts of chlorinated solvents such as occurrences, impairment of drinking water sources and detection methods for decades after the first publication in the late 1940s at a time when information was much less accessible than today. Even in later years, some publications do not distinguish between petroleum solvents and chlorinated solvents and there is no reason to believe that they refer to all types of solvents but rather to petroleum solvents. In general, there appears to have been a broad lack of awareness related to the real environmental threat posed by chlorinated solvents until the late 1970s. Only since then have we gradually become aware of the significant health and environmental risks posed by the use of chlorinated solvents and their purposeful or inadvertent release into the environment.

With the wealth of environmental occurrences, the environmental persistence and associated health risks, it is both essential and challenging to identify the source and age of chlorinated solvent spills. Environmental forensics employs techniques from different disciplines in order to do just this: to identify the source and age of environmental contamination and apportion liabilities where more than one source is involved. Usually, forensic techniques are employed in litigation; however, such techniques also permit a better understanding of subsurface contaminant history at conventional (non-litigated) sites which undoubtedly results in a better design of remediation or containment strategies. Additionally, previously unknown sources that could interfere with remediation efficiency can be identified and addressed. It is our role as forensic scientists to make the environmental community aware of such specialized forensic techniques that, beyond litigation, could save time and money in site remediation.

This chapter is written as a forensic guide, organizing the key information on chlorinated solvents in a manner useful for investigating the source and/or age of chlorinated solvents in the environment.

2   The State of Knowledge

A successful forensic investigation always starts with a comprehensive documentation on the existent state of knowledge. Data on the physical and chemical properties of chlorinated solvents are readily available in the literature, which also abounds with information on their environmental fate and transport, biodegradation and remediation efficacy. For example, a comprehensive review related to the historical context, use, amount produced, production methods and manufacturers of chlorinated solvents in the USA was provided by Doherty,[4,9] and a history of dry cleaners and solvent release sources is included in a review by Lohman.[10]

Although less abundant, environmental forensic publications are also available. Such publications focus on forensic fingerprinting techniques and their application in tracking the source or age of spilled chlorinated solvents. An excellent forensic review on chlorinated solvents was recently published by Morrison et al.[11] in a forensic textbook.[12] Other forensic review articles have been published by Morrison.[13–16] Such studies contain, apart from descriptions of forensic methods, important forensic information such as the manufacturing history of chlorinated solvents and studies of additives in selected chlorinated solvents.

Many useful publications including both reviews and forensic applications relating to chlorinated solvents (some of which are referenced within this chapter) can be found in different issues (2000–present) of the quarterly peer-reviewed journal *Environmental Forensics*, available on-line at http://www.informaworld.com/smpp/title~db=all~content=t713770863.

3   Sources and Uses

As implied by their name, chlorinated solvents effectively dissolve oil and grease. They gained widespread use largely because (compared with petroleum-based solvents) they are less flammable, less reactive and less corrosive. Chlorinated solvents gained widespread popularity and many industrial facilities have used the same or similar types of chlorinated solvents, which can make forensic identification challenging.

Based on the current knowledge, virtually all of the common industrial chlorinated solvents are of anthropogenic origin: there are no known natural occurrences of these compounds. Exceptions are chloroform and methylene chloride, which have been shown to have biogenic origins. In theory, the possibility of natural formation exists for any of the chlorinated solvents. In fact, many chlorinated organic compounds have been found to occur naturally, but none of the common industrial solvents.

Related to their application, chlorinated solvents are common industrial chemicals and are also widely used as dry cleaning chemicals. In practice, chlorinated solvents may have been present in any number of industrial operations and they are typically associated with cleaning and degreasing operations. It is their common use that makes chlorinated solvents challenging in forensic investigations. In this chapter, we briefly summarize the main sources and uses of chlorinated solvents. A thorough understanding of the main uses and potential sources for chlorinated solvents can be useful for delineating potentially responsible parties (PRPs), which is often a first step in forensic investigations.

Chlorinated solvents have had a wide variety of applications, including: dry cleaning, vapor degreasing of metal parts, hand cleaning of metal parts, paint stripping, grain fumigation, textile finishing, adhesives, fire extinguishers, replacement for PCBs in electrical transformers, carriers for rubber coatings, solvent soaps, inks, adhesives, lubricants and silicones. Chlorinated solvents have been routinely used in many industries, including rail, defense, automotive, aircraft, electronics, missiles, machinery, photographic film, semiconductors, motors, generators and appliances, boat building, shoe manufacturing, food processing and industrial paints. Chlorinated solvents were also ingredients in household cleaners and chemicals for maintenance of septic tanks. TCE was even used in dentistry as a general anesthetic and analgesic.

4 Traits and Environmental Behavior

Any forensic investigation should include a good understanding of the contaminant itself, including its structure, properties and environmental fate and transport. Such an understanding is mandatory when designing a forensic strategy, choosing appropriate forensic methods and interpreting the results. An excellent review of the environmental fate of different released organic compounds based on available physical-chemical data was published by Olsen and Davis.[17] We recommend that review for a good understanding of the principles that govern contaminant fate and transport in subsurface environments.

Table 1 gives a summary of the physical and chemical characteristics of chlorinated solvents and their subsequent effect on the environmental fate of released solvents. The exact values of different parameters (vapor pressure, solubility, boiling point, Henry's law constant, *etc.*) are not given, as they are readily available in chemistry textbooks and in material safety data sheets (MSDS) for individual products. A synthesis of the physical and chemical

Table 1   Physical-chemical characteristics of chlorinated solvents.

Characteristic/indicating parameter	Environmental effect
High volatility/vapor pressure Low sorption to soil/chemical structure and electrical charge	If released at the surface or close to surface, evaporation is an important process governing the fate of chlorinated solvents in the vadose zone; thus a portion of the mass of released solvents will evaporate, while the rest will infiltrate through the soil column with little adsorption/retardation reaching the ground-water table relatively rapidly (compared with other organic compounds)
Ability to dissolve in water/solubility	• Dissolution is expected when in contact with water; dissolved chlorinated solvents may travel with groundwater without small amount of retardation • Depending on the aquifer conditions, chlorinated solvents may not be degraded at all or, under anaerobic conditions, can be metabolized to less chlorinated daughter products (such as vinyl chloride) that often are more toxic than the parent compound • It is generally accepted that concentration of a chlorinated solvent in water exceeding 1% of its water solubility is indicative of the presence of DNAPLs (which are difficult to detect directly in the field); DNAPLs may sink into an aquifer constituting an ongoing source for dissolved phase contamination
Buoyancy relative to water/density	Pure chlorinated solvents are DNAPLs and have a tendency to sink into an aquifer, thus creating vertical contamination profiles; such plumes may easily escape detection, but their presence may be deduced from high concentrations in the dissolved phase
Low chemical reactivity/chemical structure	Chlorinated solvents usually persist in the environment (especially under aerobic conditions) for decades or more
Propensity to partition in organic phases/water–octanol partition coefficient	If a dissolved chlorinated solvent in groundwater encounters an LNAPL, it has a tendency to accumulate in the LNAPL; up to approximately 33% (by volume) of chlorinated compounds can dissolve in a typical petroleum LNAPL without breaking the phase and without becoming denser than water. When an LNAPL is present, the fate and transport of chlorinated solvents can be affected – chlorinated solvents can appear in patterns that parallel the LNAPL, misleadingly suggesting a common source
Partition into water *vs.* vapor phase/Henry's law constant	The lower the value of Henry's law constant, the higher is the amount of partitioning from

Table 1 (*Continued*).

Characteristic/indicating parameter	Environmental effect
Little flammability/flash point	water to vapor; most chlorinated solvents have a tendency to partition into the vapor phase This is one of the characteristics that made chlorinated compounds the solvents of choice to replace the more flammable petroleum solvents

properties for many chlorinated solvents is also given in a recent forensic textbook.[11] Our goal was to characterize generally the properties of chlorinated solvents as a class of chemicals and establish the general trends in environmental behavior derived from such characteristics. Predicting the behavior of chlorinated solvents released into different environments allows the interpretation of environmental data and the reconstruction of a release's evolution as it interacts with its environment. Indirectly, potential sources of chlorinated solvents may be either validated or eliminated based on such evaluation (*e.g.* proximity and duration needed to reach the actual position of the contaminant plume at a particular time).

The following paragraphs provide a summary of the environmental fate and transport of chlorinated solvents as reflected in Table 1. Chlorinated solvents are hydrocarbons that have one or more hydrogen atoms substituted by chlorine atoms. Such substitution of a very light element (hydrogen with an atomic mass of 1) with a much heavier element (chlorine with an atomic mass of 35.5) has obvious consequences, increasing the overall mass and hence the density of chlorinated compounds as compared with the corresponding hydrocarbons. Thus, pure chlorinated solvents are dense, non-aqueous phase liquids (DNAPLs) and, because they are denser than water, they will tend to sink in water, whereas the corresponding hydrocarbons are light, non-aqueous phase liquids (LNAPLs) that are less dense than water and tend to float. Additionally, the introduced chlorine atom(s) add negative charges that result in repulsion by soil or other surfaces in porous media which are usually negatively charged. This repulsion with respect to porous media particles is, in general, stronger than in the case of the corresponding petroleum hydrocarbons. In practice, once released, chlorinated solvents will percolate through soil with less adsorption and can reach groundwater faster than the corresponding hydrocarbons. Once in groundwater, chlorinated solvents tend to show lower retardation than the corresponding hydrocarbons thus migrate in groundwater at a higher rate.

Another important parameter, the high volatility of chlorinated solvents, makes evaporation an important mechanism governing the fate of chlorinated solvents in the environment. Part (but seldom all) of any chlorinated solvent release at the surface or in the unsaturated portions of the vadose zone will be transferred into gaseous form and thus escape from the soil or groundwater.

The high volatility of chlorinated solvents generated the early misconception that chlorinated solvents, if dumped on the ground, will completely vaporize rather than infiltrate into the subsurface. This misconception prevailed for many years, as illustrated by the 1972 suggestion that 'spent chlorinated solvents should be removed to a safe location (away from inhabited areas, highways, buildings or combustible structures) and poured on to dry sand, earth or ashes, then cautiously ignited',[18] a recommendation provided by the representative body for the property and casualty insurance industry in Washington, DC. Such dumping practices resulted in much of the contamination legacy we have to deal with today.

Another misconception related to chlorinated solvents that persists even today concerns their water solubility. Chlorinated solvents are often perceived as having very low solubility in water. Although it is true that chlorinated solvents are less water soluble than some other common organic contaminants (such as benzene or MTBE), they are in fact soluble enough to contaminate water at concentrations greatly in excess of the health-established limits For example, at 20 °C, the solubility of TCE in water is approximately $1100\,mg\,l^{-1}$, whereas the MCL in the USA is just $0.005\,mg\,l^{-1}$. Once dissolved in water, chlorinated solvents tend to be fairly mobile, traveling with water with little retardation. Additionally, when an LNAPL plume (gasoline or other petroleum products, for example) is encountered, chlorinated solvents tend to partition preferentially into the non-aqueous organic phase, thus accumulating in the LNAPL. Once in the LNAPL plume, the environmental fate of the chlorinated solvents will be linked to that of LNAPL and their behavior will be different from that of the dissolved phase. Thus, the presence of LNAPL should be evaluated since the fate of chlorinated solvents can be affected by such occurrences. Even with a substantial amount of chlorinated solvent, a non-aqueous petroleum liquid can remain less dense than water, and thus maintain its buoyancy. A simple calculation of the amount of chlorinated solvent that can be dissolved in LNAPL without causing the liquid to sink shows that as much as 33% of chlorinated solvents (by volume) may be dissolved in typical petroleum LNAPL without sinking. This rough estimate was made using a density of $1.5\,g\,l^{-1}$ for the chlorinated solvent and a density of $0.7\,g\,l^{-1}$ for the LNAPL. Of note for forensic investigations is that the tendency of chlorinated compounds to be sequestered in LNAPL may impede any interpretations on source areas based on underlying dissolved groundwater concentrations. This is because the highest dissolved concentrations will usually be spatially associated with the occurrence of LNAPL. Thus the chlorinated solvent source may be misleadingly confused with the LNAPL occurrence. If LNAPL is present in the subsurface, it may be useful to employ alternative forensic methods and not rely on groundwater concentration trends when investigating potential sources of chlorinated solvents.

Another important parameter to consider when evaluating the fate and transport of chlorinated solvents is the Henry's law constant (H). This constant is indirectly correlated with the propensity of a compound to partition into water *vs.* the vapor phase. Thus, the lower is H, the more a contaminant will partition into water compared with the vapor phase. According to Olsen and Davis,[17] compounds with H values $<5 \times 10^{-5}$ will partition mostly into the aqueous phase, whereas compounds with H values $> 5 \times 10^{-3}$ will partition mostly into the vapor phase. In our case, most of the chlorinated solvents have higher H values, indicating a preference for partitioning into air.

Considering all the physical-chemical characteristics of chlorinated solvents discussed above, it becomes clear why these chemicals are persistent and mobile in the environment, leaving behind a challenging legacy for environmental scientists, engineers and policy makers.

5 Significance of Environmental Forensics for Site Remediation

In dealing with the legacy of chlorinated solvents, environmental forensics plays a decisive role not only in evaluating and allocating costs to responsible parties in the framework of litigation, but also in enabling efficient design of remediation strategies: targeting sources and/or protecting sensitive downgradient receptors. It should be obvious that a thorough characterization of the operational and release history of a site (how did the contamination get there?) is a necessary component of successful long-term remediation. Unfortunately, very few remediation programs use forensic methods for site characterization or remedial technology design. It is our experience that conventional investigation strategies focus on what is there now (in terms of contamination) and often do not reliably identify all contributing sources. This may be acceptable as long as the source can be reliably identified by other means, such as the operational history of the site. However, if unidentified sources remain (including residual soil contamination), the failure to identify them can have serious implications affecting both the remediation efficiency and the duration required for remediation. For example, at underground storage tanks sites, it is not uncommon for old, rusted, undocumented underground storage tanks to be present at the site (after accidental discovery they are sometimes referred to as 'phantom tanks'). In other cases, ongoing contamination sources may be related to discharges from off-site locations that could co-mingle with local contamination and interfere with the effectiveness of the remedial program.

It is our strong belief and recommendation that many environmental investigations can benefit from a forensic component. To date, environmental forensic methods have been narrowly associated with the legal arena, due to their widespread use for litigation support. We hope that the information presented in this chapter will increase the interest in forensic studies for non-litigated environmental cases.

6 Forensic Techniques for Tracking the Source and Age of Chlorinated Solvents

A number of forensic techniques are available to investigate the source and age of chlorinated solvents in the environment. Several recently published forensic textbooks provide general information on forensic investigative and fingerprinting methods, most of which are applicable to chlorinated solvents.[13–15,19,20] Additionally, many publications on forensic methods and applications specific to chlorinated solvents are available. For example, a chapter is dedicated to chlorinated solvents in a recent forensic textbook.[11] The present chapter presents a brief summary of forensic techniques useful for chlorinated solvent sites, emphasizing the organization of this information in such a manner that it will serve as forensic guide.

A list of the forensic techniques applicable to chlorinated solvents along with guidance for their use in forensic investigations is provided in Table 2. Different applications of these techniques illustrative of their application for source delineation and age identification are also referenced in Table 2. The following paragraphs discuss the most widely used forensic methods that have relevance for sites involving chlorinated solvents in the environment.

6.1   Chemical Fingerprinting

At first glance, chlorinated solvents may seem untraceable by chemical fingerprinting methods because they are not complex mixtures of related components (such as petroleum products or PCBs or PAHs, for example). And yet, nothing could be more misleading. Different approaches may offer enough information to track the sources and age of chlorinated solvent spills. One forensic approach for fingerprinting chlorinated solvents consists of analyzing degradation or daughter products along with the parent chlorinated solvents. The specific degradation compounds along with an understanding of the major degradation pathways (biotic *vs.* abiotic; aerobic *vs.* anaerobic), are illustrated in a recent forensic article.[11] In principle, the multitude of degradation products appearing under the multitude of environmental conditions offer the possibility to characterize and track chlorinated solvent sources. Below are given a few examples illustrative of how the ratio of parent to degradation product can lead to source definition or age dating.

6.1.1   Source Investigation by Analyzing Parent vs. *Degradation Products Downgradient of the Plume.*   An example of the use of the parent to degradation product ratio for source delineation is provided by Morrison *et al.*[11] The application principle is straightforward: by taking samples downgradient of a candidate release source, expected trends in the ratio of parent to degradation products can be predicted. If the observed trends are different, the possibility of additional sources needs to be considered. Thus, by measuring the groundwater concentration ratio, [DCE]/[TCE] + [PCE], and plotting this ratio against distance downgradient from the suspected source, a declining linear relation is expected if only one source is present. If a ratio anomaly (non-linear pattern) is observed at a certain point along the plume, an additional contributing source may be indicated at that location.

6.1.2   Age Dating 1,1,1-TCA Spills Based on Degradation Products.   Another example of the use of parent *vs.* degradation products for age dating can be used for sites with releases of 1,1,1-TCA. This is the only chlorinated solvent proven to have a significant abiotic degradation pathway that occurs in nature. Thus, for 1,1,1-TCA it is possible to estimate the residence time by using the

Table 2 Forensic techniques applicable to chlorinated solvents.

Forensic method	Applications	What to look for	Application examples and references
Classical methods			
Historical document review (first step of any forensic investigation)	Confirm potential sources/spills Evaluate any available monitoring data – sometimes sufficient to answer forensic questions Identify signature chemicals that can be associated with certain sources Identify timeframe of use	Commercial availability of the particular chlorinated solvent under study Chemical applications unique to a manufacturing activity (association with signature chemicals) Specific manufacturer information – *e.g.*, booklet on equipment type (brand of equipment) and operating manuals Solvent delivery practices Unusual occurrences at the maintenance procedures Equipment locations with possible solvent loss Disposal methods of wastes/sludges – *e.g.*, sludges from regenerative filter; wastewater from dry cleaners may contain PCE and may leak through the sewer line; 'Solvent mileage' records reflecting (for dry cleaners) the weight of fabric cleaned over a specific period and the volume of solvent purchased during that time (a measure of operational efficiency) – keeping in mind the general information that between 67–1,351 pounds of fabric could be cleaned with 1 gallon of PCE Site monitoring data comprising physical-chemical and geologic analytical information available (check governmental	Comprehensive review of sources/uses, manufacturers, manufacturing processes and amounts manufactured is provided for CT, PCE, TCE and 1,1,1-TCA by Doherty[4,9] A thorough review of the history of dry cleaners and a detailed analysis of operation and maintenance, and also of dry cleaning equipment, is provided by Lohman[10] in order to evaluate the sources (causes) of releases from dry cleaning Useful guidance related to identification of chlorinated solvents based on document review has been published by Morrison[16] General application described in forensic textbooks by Murphy and Morrison[19,20] General method description and forensic applications are thoroughly reviewed in papers by Morrison,[13–15] some of which contain special

Table 2 (*Continued*).

Forensic method	Applications	What to look for	Application examples and references
		agencies files such as Water Board or DTSC, *etc.*) Contaminants used at the site through consulting agency's files such as fire department or city/municipality files reported and suspected spills/accidents or at the site and surrounding sites historical aerial photographs showing changes in land use, structures, *etc.*	reference to chlorinated solvent identification and production history
Chemical fingerprinting: ratio of parent to degradation products	Age dating based on concentration of degradation products – only for 1,1,1-TCA plumes that degrade chemically Source identification by the ratio of parent to degradation compounds	Concentration of parent and degradation compounds Spatial distribution with aquifer, well type and land use Vertical profile in soil Concentrations in saturated *vs.* unsaturated zone	Age dating groundwater 1,1,1-TCA plumes based on the ratio of 1,1-DCE to 1,1,1-TCA is described by Murphy and co-workers[11,21]
Stabilizers/additives	Source identification by association with other ions and potentially signature compounds (*e.g.* additives, stabilizers), and also by evaluating spatial distribution	Concentrations of different original solvents correlated to their vapor pressures → estimate enrichment factor Concentration of additives and signature chemicals if identified in the sample	A list of additives (including acid inhibitors, metal inhibitors, antioxidants and light inhibitors) is given for main chlorinated solvents in a review by Morrison *et al.*[11]
Signature chemicals (other compounds associated with suspected sources)	Evaluation of accumulation of less volatile compounds during degreasing operations	Ratio analysis for age dating and identification of additional sources downgradient the main one (for example representing [DCE]/[TCE]+[PCE] *vs.* downgradient distance from the suspected source	The computation of the accumulation of PCE in spent TCE solvent is given by Morrison *et al.*[11]
Isotopic fingerprinting	Distinguish between manufacturers Link spill/plumes to potential sources	Stable isotopic composition for chlorine (^{37}Cl) – seems constant between manufacturer	An excellent review of the use of isotopic analysis in chlorinated solvent forensics

Distinguish between manufactured TCE *vs.* TCE resulting through biodegradation of PCE

Evaluate biodegradation occurrence and predict its effects on isotopic fractionation

Distinguish between biogenic and anthropogenic sources, for those chlorinated organic compounds with proven both biogenic and anthropogenic origin (*e.g.* methyl chloride and chloroform)

batches and specific to the manufacturing process

A combination of both C and Cl stable isotopic composition may be indicative for sources (both environmental samples and potential sources should be analyzed and compared)

Stable H (^{2}H) isotopic composition of TCE used to distinguish manufactured *vs.* product of degradation for TCE

Plotting the isotopic values against each others provides more accurate indication of the source

Use of both the compound specific isotopic analysis (CSIA) and compound specific radiocarbon analysis (CSRA) – for solvents such as methyl chloride and chloroform and; the biogenic compounds will have ^{14}C (radioactive C) signature associated with living organisms, while the anthropogenic compounds will not have any measurable radioactive carbon

is provided by Morrison *et al.*[11] The 1995 study of Van Warmerdan *et al.*[27] is one of the first studies on chlorinated solvent isotopic composition that demonstrates that the combination of C and Cl isotopic analysis may indicate sources and delineate contamination episodes; samples from different main manufacturers had distinct isotopic signatures

A later study[28] reports the consistency of stable Cl isotopic compositions between batches of manufacturers for both PCE and TCE; additionally, the combination of C and Cl isotopic composition may be used to study the hydrogeology of released chlorinated solvents

A 2001 study[29] further confirmed the use of stable C and Cl isotopes to distinguish between manufacturers; the study also confirmed that Cl isotopic compositions are highly fractionated during organic synthesis and may

Table 2 (*Continued*).

Forensic method	Applications	What to look for	Application examples and references
			therefore be preserved between different batches In addition to stable C and Cl isotope measurements, stable H isotope measurements provide additional information for accurately source tracking of chlorinated solvents; stable H isotopic composition may be used by itself to distinguish between manufactured TCE and TCE produced by biodegradation of PCE, as shown by Shouakar-Stash et al.[30] Apart from delineating the pollution source, the combination of classical isotope tools (^{18}O, ^{2}H and ^{3}H) with compound-specific isotopic analysis (isotopic composition of each individual solvent) allow one to analyze flow patterns and geochemical reactions, as

Groundwater modeling	Determine upgradient areas and timing for chlorinated solvent plume to reach the sampling point from different source areas Indirect determination of chlorinated solvent plume age by modeling contaminant transport (through soil	Groundwater flow directions and velocities in the study area – timing for contamination from potential sources to reach the sampling area Geology/lithology of the study area Historical monitoring data on parent compounds and degradation products (in the case of 1,1,1-TCA these products can be	General applications are described in the forensic textbooks of Murphy and Morrison[19,20] Contaminant transport models (including transport through pavement and soil as well as inverse and phase-separated

demonstrated by Klopmann et al.[31]

Stable isotope analysis of chlorinated solvents could be used to evaluate behavior in contaminated aquifers, more specifically biodegradation vs. evaporation or other types of transformation of chlorinated solvents – as shown in a study by Sturchio et al.[23]

The use of stable isotopes to evaluate natural attenuation of chlorinated solvents is also proven in a study by Sturchio et al.[32]

Other studies prove the potential to use isotopic fractionation to evaluate the environmental behavior of chlorinated solvents (e.g. the study of Heraty et al.[33] for dichloromethane)

Table 2 (*Continued*).

Forensic method	Applications	What to look for	Application examples and references
	and groundwater) from potential source(s) Age-dating based on the position of plume front	used to estimated groundwater velocities by hydrochemical facies method (plotting mole fraction data on a ternary diagram) When using the position of the plume front, the following parameters should be considered: groundwater velocities, length of the plume, longitudinal dispersion, hydraulic conductivity, porosity and retardation factor → solvent transport velocities	groundwater models) are reviewed by Morrison[13] Degradation models available for chlorinated solvents are reviewed by Morrison[14] A formula for calculating the age of chlorinated solvent plume based on the position of the plume front is given by Morrison et al.[11]
Statistical evaluation	Identify and delineate sources Compare the behavior of a particular plume with a population of plumes from other sites that may be influenced by similar mechanisms of transformation processes (or testing plume behavior)	Chemical composition of samples – used in the statistical calculations (entered parameters) → association (cluster) of samples based on chemical similarity Similarity of environmental conditions to which different samples have been exposed – mandatory for valid statistical applications	General applications are described in Murphy and Morrison[19,20] Discerning trends in chlorinated hydrocarbon plume behavior from a diverse multi-site data set was proven efficient in chlorinated solvent forensics in the study of McNab[34]
Emerging applications			
Indirect age-dating (minimum groundwater age)	Determine the recharge time of groundwater that will indirectly represent the minimum time since the chlorinated compound entered the groundwater body – assuming that contamination is in contact and reaches groundwater through rainwater or surface water infiltration	Measure groundwater age by the concentration of a series of atmospheric contaminants such as chlorofluorocarbons and tritium isotopes (He measurements are also needed) Groundwater samples should be collected from the vertical or horizontal leading edge of the plume	General use of atmospheric contaminants to age date groundwater releases IS was reviewed by Oudijk[35] and illustrated by several successful applications involving chlorinated solvents

Technique	Objectives	Required data / parameters	Comments
		The screen length of the well used for groundwater sampling should be minimal and within a discrete hydrogeological zone The laboratory should use a method with a very low detection limit (*e.g.* $0.1\ \mu g\,l^{-1}$ or less) The presence of natural or artificially induced radiation in the study area needs to be considered if tritium isotopes are used for groundwater age dating Additional (other than atmosphere) sources of atmospheric contaminants used to track groundwater age should be evaluated	The specific application of the method (using CFCs and tritium) to estimate the minimum age of chlorinated solvent plumes was also published by Oudijk;[36] this study presents the values of historical CFC concentrations in atmospheric water at different temperatures; such values could be used by comparison with study values; age dating groundwater based on tritium and helium concentrations is given by the following equation:[36] $T =$ $$T = (T_{1/2}\ln 2)\ln(1 + {}^3He_{tri}/{}^3H)$$ where T = age of water, $T_{1/2}$ = half-life of ${}^3H$ (12.43 years) and He_{re} = concentration of radiogenic He (formed from ${}^3H$)
Dendroecology (using tree ring data to age date and characterize environmental releases)	Age-date chlorinated solvent releases with precision to the year, even when contamination events happened a long time ago (100 years or more), providing trees of appropriate age are present Source delineation through proven travel pathways Characterize the type of release: sudden and accidental *vs.* gradual leak	Ring width anomalies Cl concentration peaks along the tree core sample Associated elemental peaks (Na, K, *etc.*) – to evaluate other possible contributions resulting in Cl elemental peak in the study tree (such as Cl from salt used for de-icing) Statistical data on ring width for similar species in the study area (for comparison)	The method is based on the fact that chlorinated solvents leave traces in trees – studies in phytoremediation established the absorption of chlorinated solvents in trees and their presence (unmodified) in different tree parts (including the trunk) – for example, the study of

Table 2 (*Continued*).

Forensic method	Applications	What to look for	Application examples and references
	Identify multiple releases	Climatic data for the study area – evaluate potential climatic impact on tree growth	Chard et al.[3] demonstrates the presence of TCE and its metabolites in trees exposed to different concentrations of TCE General information on the technique and successful applications (including chlorinated solvents) are presented in recent publications by Balouet et al.[25,26]

chemical degradation pathway. The ratio of 1,1-dichloroethene (1,1-DCE) to 1,1,1-TCA is used to date a release based on the rate of chemical hydrolysis. A comprehensive study related to the use of this method to age date groundwater plumes has been published by Gauthier and Murphy.[21] The chemistry of the hydrolysis process consists in the following reactions:

$$CH_3CCl_3 \rightarrow CH_3C^+Cl_2 + Cl^- \quad \text{rate} - \text{limiting step}$$

Two subsequent pathways (in parallel) follow:

$$CH_3C^+Cl_2 + Cl^- + 2H_2O \rightarrow CH_3COOH + 3HCl \quad \text{rate constant} = k_s$$
$$CH_3C^+Cl_2 + Cl^- \rightarrow CH_2CCl_2 + HCl \quad \text{rate constant} = k_e$$

The total abiotic degradation rate constant (k) is equal to the sum of the individual rate constants from the reactions described before (substitution and elimination):

$$k = k_s + k_e$$

Based on the measurement of degradation products, there are two approaches for calculating the residence time (t) for age dating 1,1,1-TCA releases:[21]

Method 1 calculates t based on single measurements of 1,1,1-TCA and 1,1-DCE concentrations and is based on laboratory data related to the rate constant values (k and k/k_e) at a known groundwater temperature:

$$t = 3.63 \times 10^{-22} \exp[54.065/(1 + 3.66 \times 10^{-3} T)] \ln\{1 + 4.76([DCE]/[TCA])\}$$

where T = temperature (°F), [DCE] and [TCA] are concentrations (mg l^{-1}) and t = time (years).

Method 2 calculates t based on time-series measurements of DCE and TCA concentrations, by plotting $\ln(1 + k[DCE]/k_e[TCA])$ as a function of time. This function plots as a line that intersects the x-axis at $t = 0$ or the time of the release.

Different factors need to be carefully considered and evaluated before the use of this method:

• Processes that may transform and/or produce more 1,1-DCE will interfere with the 1,1-DCE produced by degradation of 1,1,1-TCA.
• Other degradation pathways for 1,1,1-TCA could play a more or less important role on a case-by-case basis. Such pathways should be evaluated and eliminated for the age dating method presented previously to be valid.
• Groundwater conditions and sorption to soil and sediments could also affect age dating based on chemical degradation of 1,1,1-TCA.
• The presence of DNAPL or LNAPL can influence the interpretation of an age dating investigation;

Uncertainties related to the age dating method should be considered and evaluated in interpretations:

- Samples should be collected at wells located in the closest area to the plume front with the highest contaminant concentrations.
- Steps should be taken to determine if the groundwater temperature at the time of the study is representative over the period of interest.

6.1.3   Use of Signature Chemicals for Source Identification.   Some commercial chlorinated solvents contain stabilizers to enhance shelf-life or minimize reactions with metals during use. The stabilizers added to chlorinated solvents include 1,4-dioxane, 1,3-dioxolane, 1,2-butylene oxide, *N*-methylpyrrole, methyl ethyl ketone (MEK), ethyl acetate, acrylonitrile, nitromethane, dialkyl sulfoxides, sulfides, sulfites, tetraethyllead, isopropyl alcohol, *tert-* and *sec*-butyl alcohol, morpholine, tetrahydrofuran and toluene. Of the different chlorinated solvents, 1,1,1-TCA had the highest occurrence and highest concentrations of stabilizers. Other additives may have been added to certain chlorinated solvents based on the exact type of application. The manufacturer's operating manual often contains valuable information related to the use of particular chemicals as additives to chlorinated solvents for particular applications. Whatever the case, if information is available on the additives or stabilizers blended with the particular chlorinated solvent used at a site, such compounds can be tested in environmental samples, and, if present, may provide compelling forensic evidence for source identification.

6.2   Isotopic Fingerprinting

Along with chemical fingerprinting discussed above, isotopic fingerprinting is an essential technique used to track the source and age of chlorinated solvent spills. There are different ways of measuring isotopes, including:

- bulk isotopic measurements of the isotopic composition of a certain element (integrating all compounds that may be present in an environmental sample);
- compound-specific stable isotopic analysis (CSIA) measuring the stable isotopic composition of individual compounds extracted from an environmental sample;
- compound-specific radioactive carbon analysis (CSRA) measuring the radioactive carbon isotopic composition of extracted individual compounds.

The bulk isotopic composition may be used to compare and cluster experimental samples. Different types of CSIA isotopic measurements (of both carbon and chlorine isotopes), as shown in Table 2, have the potential for source delineation and for identifying original manufacturers. Additionally, CSIA for C isotopes is a powerful tool to study the environmental fate and transport of released chlorinated solvents, and also for distinguishing between bio-degradation and evaporation processes. Available data suggest that chlorine isotopes may remain constant between one manufacturer's batches but may differ between manufacturers, and therefore can be used as a signature for identifying original manufacturers. CSIA for hydrogen isotopes is a powerful tool for distinguishing between manufactured TCE and TCE produced by biodegradation of PCE. This is due to the large range of hydrogen isotopic values because TCE formed from PCE bears the H atom from surrounding environmental water, whereas manufactured TCE is the result of a series of chemical reactions resulting in very positive isotopic values for H. In general, a combination of isotopic data for more than one element in a compound (*e.g.* C, Cl and H for chlorinated solvents) is recommended for reliable forensic determinations.

Apart from the direct forensic use of CSIA in evaluating the source of chlorinated solvents in the environment, CSIA may be used to evaluate the degradation pathways of released solvents, especially to prove biodegradation. This is because isotopic fractionation occurs during biodegradation but significant fractionation apparently does not occur from most other physical-chemical processes contributing to the fate and transport of chlorinated compounds in the environment. Biodegradation preferentially consumes the lighter carbon isotopes, resulting in enrichment of the heavier isotope in residual contamination. This isotopic fractionation associated with bio-degradation is distinct from the fractionation produced by another type of important environmental transformation: volatilization. A distinct isotopic effect is produced due to volatilization, theoretically making it distinguishable from biodegradation. Huang *et al.*[22] showed a negative correlation between the stable carbon and hydrogen isotopic composition of TCE and dichloromethane (DCM). In other words, a decrease in the heavier isotope of both C and H was observed with volatilization.[22] This observation was confirmed by a more recent study[23] that observed a negative correlation of carbon isotopic composition with respect to degree of evaporation, for most of the solvents studied, whereas the chlorine isotopic composition maintained a positive correlation similar to that produced by biodegradation. In spite of the scarce data in this respect, it appears that distinct isotopic effects are associated with bio-degradation *vs.* evaporation. There is limited information related to the extent, if any, of isotopic fractionation associated with other attenuative processes such as diffusion and dispersion. Until further research improves our understanding of isotopic behavior during these complex processes, it may be inferred that an increase in carbon isotopic composition in chlorinated solvents is correlated with the extent of biodegradation that the chemical has experienced. As with any environmental testing, it is important to adhere to appropriate quality control protocols in the laboratory.[24]

The CSIA analysis of ^{14}C (radioactive carbon) is a technique that can be used to distinguish between biogenic and anthropogenic sources of chlorinated organic compounds. This is because radioactive carbon is absorbed by living plants from the atmosphere through photosynthesis and is totally absent in manufactured compounds. This radioactive isotopic method has forensic value in applications involving the chlorinated solvents known to have both biogenic and anthropogenic origin. Such solvents are scarce and include methyl chloride and chloroform. Interestingly, an abundance of naturally occurring chlorinated organic compounds have been isolated (some 2200 at the last count) but, as noted above, very few of the common industrial solvents have ever been found to occur naturally.

6.3   Dendroecology

Dendroecology is an emerging forensic application of well-established procedures that uses tree ring data to age date and characterize environmental releases. It is the only forensic method that can estimate the age of contamination with precision to the year. The method also has the advantage of being independent of the physical presence of contamination at the time of sampling. The method is based on well-established dendrochronology principles that link the width of the yearly tree growth rings with climate and any other environmental factors that could affect the health of the tree. The forensic application consists of evaluating the effect of contamination on ring width and on the chemical composition of each growth ring. When contamination is present at the root zone, it will enter the tree with the sap (liquid that circulates through tree), some contamination will translocate into the trunk where it may leave a permanent chemical trace in the wooden growth rings. Ideally, environmental insults will be measurable in both the growth of the tree (reflected though the width of the rings) and the chemical composition of growth ring. A peak of certain element(s) associated with a particular contaminant in a growth ring followed by the reduction in width of the same ring or of rings from consecutive years (after contamination event) provides evidence of the exact timing of the arrival of contamination into the tree root environment. For example, compounds containing chlorine (such as chlorinated solvents), once absorbed by the tree, can cause the appearance of a chlorine anomaly within the tree growth ring corresponding to the year when contaminant first entered the tree through the root system. The method can also be used for petroleum hydrocarbons (such as diesel and fuel oil) using sulfur as an elemental marker. Obviously, variables such as distance from the source and depth to groundwater have important influences on the strength of the tree ring signal.

    Balouet and Petrisor (the first of Environmental International, Paris) have been involved in successful applications of the dendroecology method to age date and characterize chlorinated solvent plumes.[25] More information on the method and its application can be found in several pioneering publications.[25,26] The method is of particular interest for environments with shallow groundwater (above 30 ft below ground surface) and at sites where old trees are present downgradient of potential source(s). Apart from the high precision, the method has the advantages of being environmental friendly, easy to apply (simple sampling device) and relatively inexpensive.

7 Conclusions and Perspectives

More than a century of handling chlorinated solvents combined with the lack of environmental awareness (until recent decades) resulted in a contamination legacy that will continue to persist for years to come. The challenge of this legacy involves both remediation and cost recovery from responsible parties. Environmental forensics uses a variety of techniques to investigate the source and age of contamination and allocate responsibilities and is commonly used in litigation. However, forensic methods are also valuable for site characterization and the selection and design of appropriate remediation systems. This chapter is written as a forensic guide reviewing key information of relevance to sites with chlorinated solvent releases and also investigative methods for tracking chlorinated solvents in the environment. Several forensic methods have been identified for their deductive power and flexibility, including chemical and isotopic fingerprinting, in addition to dendroecology. We hope that this chapter will encourage the use of forensic techniques in site characterization and remedial design, and also encourage the development of new forensic methods. It is the role of forensic scientists to recognize novel uses of existing methods from other disciplines and to develop new forensic applications to solve complex environmental problems. Dendroecology is just one example of how powerful such applications can be in the service of environmental science.

References

1. P. J. Squillace, M. L. Moran, W. W. Lapham, C. V. Price, R. M. Clawges and J. S. Zogorski, *Environ. Sci. Technol.*, 1999, **33**, 4176–4187.
2. US Environmental Protection Agency, *Contract Laboratory Program Statistical Database*, EPA, Washington, DC, 1989.
3. B. K. Chard, W. J. Doucette, J. K. Chard, B. Bugbee and K. Gorder, *Environ. Sci. Technol.*, 2006, **40**, 4788–4793.
4. R. E. Doherty, *Environ. Forensics*, 2000, **1**, 69–81.
5. F. A. Lyne and T. McLachlan, *Analyst*, 1949, **93**, 513.
6. M. O. Rivett, S. Feenstra, L. Clark, F. A. Lyne and T. McLachlan, *Environ. Forensics*, 2006, **7**, 313–323.
7. S. Amter and B. Ross, *Environ. Forensics*, 2001, **2**, 179–184.
8. R. E. Jackson, *Environ. Forensics*, 2003, **4**, 3–9.
9. R. E. Doherty, *Environ. Forensics*, 2000, **1**, 83–93.
10. J. H. Lohman, *Environ. Forensics*, 2002, **3**, 35–58.
11. R. D. Morrison, B. L. Murphy and R. E. Doherty, in *Environmental Forensics. Contaminant Specific Guide,* ed. R. D. Morrison and B. L. Murphy, Elsevier Academic Press, 2006, pp. 259–277.
12. R. D. Morrison and B. L. Murphy (eds.), *Environmental Forensics. Contaminant Specific Guide*, Elsevier Academic Press, 2006.
13. R. D. Morrison, *Environ. Forensics*, 2000, **1**, 131–153.
14. R. D. Morrison, *Environ. Forensics*, 2000, **1**, 157–173.
15. R. D. Morrison, *Environ. Forensics*, 2000, **1**, 175–195.

16. R. D. Morrison, *Environ. Claims J.*, 2001, **13**, 95–104.
17. R. L. Olsen and A. Davis, *Hazard. Mater. Control*, 1990, **3**(3), 39–64.
18. American Insurance Association (AIA), *Chemical Hazards Bulletin*, C-86, AIA, New York, 1972, p. 41.
19. B. L. Murphy and R. D. Morrison (eds.), *Introduction to Environmental Forensics*, Elsevier Academic Press, 2002.
20. B. L. Murphy and R. D. Morrison (eds.), *Introduction to Environmental Forensics*, Elsevier Academic Press, 2nd edn., 2007.
21. T. D. Gauthier and B. L. Murphy, *Environ. Forensics*, 2003, **4**, 205–213.
22. L. Huang, N. C. Sturchio, T. Abrajano Jr., L. J. Heraty and B. D. Holt, *Org. Geochem.*, 1999, **30**, 777–785.
23. N. C. Sturchio, L. Heraty, B. D. Holt, L. Huang, T. Abrajano and G. Smith, in *Proceedings of the 2nd International Conference on Remediation of Chlorinated Recalcitrant Compounds*, ed. G. B. Wickramanayake, A. R. Gavaskar and M. E. Kelley, Battelle Press, Columbus, OH, 2000.
24. B. D. Holt, L. J. Heraty and N. C. Sturchio, *Environ. Pollut.*, 2001, **113**, 263–269.
25. J. C. Balouet, G. Oudijk, K. Smith, I. G. Petrisor, H. Grudd and B. Stocklassa, *Environ. Forensics*, 2007, **8**, 1–17.
26. J. C. Balouet, G. Oudijk, I. G. Petrisor and R. Morrison, in *Introduction to Environmental Forensics*, 2nd edn., ed. B. L. Murphy and R. D. Morrison, Elsevier, Amsterdam, 2007, pp. 699–731.
27. E. M. van Warmerdam, S. K. Frape, R. Aravena, R. J. Drimmie, H. Flatt and J. A. Cherry, *Appl. Geochem.*, 1995, **10**, 547–552.
28. K. M. Beneteau, R. Aravena and S. K. Frape, *Org. Geochem.*, 1999, **30**, 739–753.
29. N. Jendrzejewski, H. G. M. Eggenkamp and M. L. Coleman, *Appl. Geochem.*, 2001, **16**, 1021–1031.
30. Shouakar-Stash, S. K. Frape and R. J. Drimmie, *J. Contam. Hydrol.*, 2003, **60**, 211–228.
31. W. Kloppmann, D. Hunkeler, R. Aravena, P. Elsass and D. Widory, *Geophys. Res. Abstr.*, 2005, **7**, 08004.
32. N. C. Sturchio, J. L. Clausen, L. J. Heraty, L. Huang, B. D. Holt and T. A. Abrajano Jr., *Environ. Sci. Technol.*, 1998, **32**, 3037–3042.
33. L. J. Heraty, M. E. Fuller, L. Huang, T. Abrajano Jr. and N. C. Sturchio, *Org. Geochem.*, 1999, **30**, 793–799.
34. W. W. McNab Jr., *Environ. Forensics*, 2001, **2**, 313–320.
35. G. Oudijk, *Environ. Forensics*, 2005, **6**, 345–354.
36. G. Oudijk, *Environ. Forensics*, 2003, **4**, 81–88.

Groundwater Pollution: The Emerging Role of Environmental Forensics

STANLEY FEENSTRA AND MICHAEL O. RIVETT

1 Introduction

In recent years, subsurface forensics has emerged as a definable subject area in the field of contaminant hydrogeology or groundwater pollution. For the purpose of this chapter, a definition modified from Morrison and Murphy[1] will be employed. Subsurface forensics will refer to: '*the systematic and scientific evaluation of physical, chemical and historical information for the purpose of developing defensible scientific and legal conclusions regarding the source or age of contaminant release to the subsurface environment*'. The conclusions of subsurface forensic investigations are used in legal proceedings in either the courts or administrative tribunals. Subsurface environmental forensics has been a focus of recent books by Morrison,[2] Sullivan *et al.*,[3] Murphy and Morrison, 1st edition[4] and 2nd edition,[5] and Morrison and Murphy,[1] and of the journal *Environmental Forensics*, which began publication in 2000.

Subsurface forensics may include elements such as assessment of site history, collection of field data, physical or chemical analysis or methods of data presentation and interpretation. Many such work elements are part also of regular contaminant hydrogeological studies intended for the characterization of sites, evaluation of contaminant fate and transport and the design of remedial measures to manage risks posed by groundwater pollution. However, subsurface forensic methods are engaged in combination with regular contaminant hydrogeology studies to answer different questions. Legal proceedings related to subsurface contamination commonly seek to determine responsibility for the investigation and clean-up of contaminated sites or allocation of responsibility of multiple parties to assign investigation and clean-up costs. Responsibility may lie with former or present site owners, suppliers of equipment or chemicals or insurers. The process of allocation of responsibility differs internationally due to the varying national legislation in place and case law established. Typically it is a complex, protracted and expensive process for many sites.

Issues in Environmental Science and Technology, No. 26
Environmental Forensics
Edited by RE Hester and RM Harrison
© Royal Society of Chemistry 2008

Almost all legal proceedings are adversarial in nature and the conclusions put forth by expert witnesses for one side will be scrutinized and challenged by experts for the opposing side. In most cases, the degree of scientific review and scrutiny in legal proceedings far exceeds that applied to regular hydrogeological studies and in some cases exceeds that of normal scientific peer review in academia. As a result, the questions posed and answers expected of expert witnesses regarding subsurface contamination often span a broad spectrum from the most basic of hydrogeological principles to esoteric aspects of analytical chemistry.

The basic questions to be answered in the majority of subsurface forensic studies are broadly similar to those posed in the classical detective board game *Clue*: 1. WHO DONE IT?, ... 2. WHERE? ... 3. HOW?

Within each of these basic questions may be a multitude of more specific and more detailed questions depending on the site conditions and the issues of concern in a particular case. The forensics methods likely recognized by most readers are chemical or isotopic analysis techniques that attempt to identify or distinguish between different sources of contamination ('fingerprinting') or attempt to determine when the contaminants may have been released to the subsurface ('age dating'). Many of the fingerprinting methods for petroleum hydrocarbons have their beginnings in geochemical studies of crude oil formation and prospecting and environmental studies of marine oil spills. The other chapters in this volume describe many of these techniques. However, in general, the results of such 'specialist' analytical methods alone do not provide answers to the basic or even more specific questions. Various elements of more routine contaminant hydrogeological studies, assessment of site history and knowledge of chemical uses and practices must be combined with specialist analytical methods.

This chapter gives examples of some of the important or challenging questions commonly posed in subsurface forensic cases and notes both the hydrogeological and forensic methods that may contribute to providing useful answers. Examples given here are based on the experiences of the authors with specific cases, which must remain unidentified. Because chlorinated solvents and petroleum hydrocarbons comprise some of the most widespread of subsurface chemical contaminants,[6,7] many of the examples given here will relate to releases and groundwater contamination by these types of chemicals.

2   The Fundamental Questions

A typical circumstance at a contaminated site involves some past release of chemical materials to the subsurface which has caused contamination of the soil or groundwater. In some cases, releases may have occurred many decades before the consequences became apparent. Many sites are complex: releases may have occurred of various chemicals, at various locations, at various times, by various potentially responsible parties. Significant concerns typically exist surrounding the longevity of ongoing sources of contamination present within the subsurface, particularly where light or dense non-aqueous phase liquid (LNAPL, DNAPL) sources reside within aquifers that slowly dissolve over decade or even century time-scales.[8,9] Most commonly, however, the

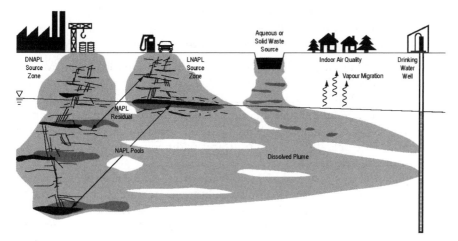

Figure 1 Conceptual models of groundwater pollution arising from DNAPL and LNAPL source zones and aqueous or solid waste sources.

environmental concern or liability primarily arises from the dissolved-phase plumes of groundwater contamination that have migrated away from the release area source zones. Plumes may potentially cause contamination beyond the site boundary of the wider groundwater resource, specific supply wells and surface waters or wetlands to which groundwater may naturally discharge.

Our conceptual models for understanding groundwater pollution are therefore important; some aspects of current models are illustrated in Figure 1. The figure importantly incorporates the DNAPL paradigm applicable to the transport of immiscible dense organic liquids; this conceptualization came to be widely accepted by practitioners only from around the 1980s onwards;[10] hitherto it was not widely appreciated that such liquids could rapidly and deeply penetrate the subsurface as immiscible DNAPLs far below the water table, subsequently to dissolve slowly in flowing groundwater at depth. Such conceptual models illustrate the importance of specific consideration of both the release chemical type and the subsurface nature. They provide the contaminant hydrogeological framework fundamental to the effective application of environmental forensics to groundwater pollution.

The basic questions of Who?, Where? and How? need to be rearranged and expanded to relate properly to the conceptual model of a contaminated site which is the subject of litigation.

The question Where? commonly expands more specifically to:

• On which site or area of a site, did the releases occur?
• Where are any persistent source zones?

The question How? commonly expands to:

• Were the releases from leaks and spills, intentional disposal, leaking sewers?

- Were there releases of different chemicals?
- Did releases occur at different times?
- Can the date of the releases be determined?

To answer the question Who?, often it is usually necessary first to answer the questions Where? and How? and combine the answers with specific knowledge of the history of the site or at least general knowledge of the activities at the site. It must be noted that the specific questions posed to a forensic investigator may vary widely from case to case depending on the contaminants of concern, the nature of the site operations, the subsurface conditions at the site and the legal issues of the specific case.

A supplementary question posed to forensic investigators in some cases relates to the historical knowledge regarding subsurface contamination by specific chemicals. For example, when did it come to be known that releases or disposal of a specific chemical or group of chemicals could cause subsurface contamination problems? Alternatively phrased, at what date could the contamination problem have been reasonably foreseen?[11] For example, if the releases or disposal took place at a time when the resultant contamination could have been reasonably expected, the responsible party might be denied insurance coverage to pay for the investigation and clean-up or may be charged in civil litigation with punitive damages in addition to cost of true damages caused by the contamination.

3   Subsurface Forensic Methods

Two fundamental types of information are employed in subsurface forensics studies of contaminated sites: existing documentary evidence and measurements and sample analysis data. Documentary evidence includes site-specific information about the contaminated site, or 'site history', and also general information or knowledge that may be related to the activities at the site.

3.1   Site History

Murphy and Morrison[1] and Sullivan et al.[3] gave detailed descriptions of the types of information and the sources of such information that can contribute to an understanding of site history. Company documents such as plant engineering drawings, chemical purchase records and interviews with current and former employees, regulators or chemical suppliers can give information on the types and quantities of chemicals used and the nature of chemical storage, handling and distribution practices. Archival aerial photographs can provide information on historical outdoor chemical storage or waste disposal activities.

However, in many circumstances, site-specific information may be lacking and forensic studies must be supplemented by general technical or engineering knowledge regarding the uses of specific chemicals or the nature of specific industrial operations. Morrison and Murphy[1] compiled a multitude of information relevant to subsurface forensic studies on some of the most common chemical contaminants which included mercury, lead, chromium, radioactive compounds, pesticides, perchlorate, polychlorinated biphenyls (PCBs), chlorinated solvents, arsenic, dioxins and furans, polycyclic aromatic hydrocarbons (PAHs), gasoline and crude oil and other refined hydrocarbon products. For each of these contaminants, the principal industrial uses were described. Among other more detailed presentations of general knowledge of chemical uses or industrial processes have been those of Doherty[12,13] on carbon tetrachloride, perchloroethene, trichloroethene and 1,1,1-trichloroethane, Gibbs[14] on gasoline additives, Lohman[15] on the dry cleaning industry and Hamper[16] on manufactured gas plants. The Toxicological Profile Reports published by the Agency of Toxic Substances and Disease Registry of the US Department of Health and Human Services can also provide general information on the uses of various industrial chemicals. It is almost always preferable for forensic experts to refer to such published sources of general information to the greatest extent possible to supplement their personal and/or general knowledge and experience of chemical uses and industrial operations.

3.2 Site Testing

Site testing includes all the measurements and sample analysis data collected as part of a subsurface forensics study. In the many cases, site testing data used by forensic investigators include only work that has been already collected as part of 'routine' hydrogeological studies. 'Specialist' sample analysis methods, data evaluation or numerical modelling techniques are not commonly used in routine hydrogeological studies. They are used to answer more specific forensic questions but in almost all cases such specialist methods rely heavily on a foundation of the routine hydrogeological studies.

It is also the nature of litigation matters that interpretations and opinions must sometimes be developed without all the routine hydrogeological or specialist analysis methods that might be applicable. Such data might be unavailable because of access restrictions to the site, lack of funding or lack of time. In addition, counsel acting in such matters may be averse to using methods whose effectiveness cannot be predicted with confidence or that have not been documented in peer-reviewed scientific literature or that may potentially yield results directly favourable to the opposing side.

3.2.1 Routine Testing. The data include basic hydrogeological data on the: site geology, hydrostratigraphy, groundwater flow direction and flow velocity and spatial distribution and concentrations in soil and groundwater of the contaminants of concern. Relevant data may be available also to assess the rates of contaminant migration and attenuation processes (*i.e.* sorption coefficients, degradation half-lives). Some aspects of routine hydrogeological studies take on greater significance when the data are employed in forensic analyses.

For example, the direction of groundwater flow is often an important factor in assessing the origin of a groundwater plume, particularly in an urban area where there may be a number of potential sources. The confidence with which forensic conclusions can be drawn on the direction of groundwater flow will depend on: the number, spacing and vertical intervals of the monitoring wells, the frequency and duration of groundwater level monitoring and the accuracy of groundwater level measurements and reference point elevations. However, in some cases, forensic investigators must also interpret groundwater elevations and flow directions for times that may be decades before the period of actual monitoring. In such cases, typically with a paucity of available data, interpretations must consider the effect on groundwater elevations or flow directions due to historic groundwater pumping, long-term climatic effects such as droughts and short-term climatic events such as floods or hurricanes.

Routine hydrogeological investigations of contaminated sites typically utilize methods for sampling and chemical analysis of soil, soil gas and groundwater that have been standardized and approved by environmental regulatory agencies such as the US Environmental Protection Agency (EPA)[17] or relevant scientific organizations such as the American Public Health Association (APHA).[18] These routine methods include analyses for inorganic constituents such as major cations, anions and nutrients, toxic metals, volatile organic compounds (VOCs), semi-volatile organic compounds (SVOCs), organic chemical pesticides, PAHs and PCBs. It is important to note that most routine methods were developed in the 1970-80s and thereby facilitated the general onset of awareness of subsurface contamination from around the mid-1970s onwards.[19] As such, valuable historical water quality data on contaminants of concern are rarely available prior to the mid-1970s to early 1980s or even later. This particularly applies to organic contaminants, often key risk drivers at many sites.

Most of these routine methods are sufficiently reliable to provide useful data for forensic investigations. However, there may be circumstances where routine methods need to be enhanced. For example, if the contaminants of concern may be affecting drinking water supply wells at relatively low concentrations, it may be desirable to use analytical methods capable of lower detection limits or it may be desirable to collect a greater number than normal of quality assurance/quality control (QA/QC) samples such as trip blanks, field blanks and field duplicates to confirm that the analysis results are reliable and accurate.

3.2.2 Specialist Methods. The specialist methods employed in subsurface forensic studies can be categorized as sample analysis methods and modelling methods. Details of many of these methods and their applications were presented by Murphy and Morrison[4,5] and in other chapters in this volume.

3.2.2.1 Sample Analysis Methods. Most specialist sample analysis methods used in subsurface forensics aim to identify or distinguish between different sources of contamination (source evaluation) on the basis of: the presence of specific contaminants or associated compounds, the concentrations, ratios or chromatographic patterns of assemblages of organic compounds or the stable isotopic composition of one or more of the compounds. Collectively, the general applications of such types of analysis methods are referred to as fingerprinting. High-resolution gas chromatography–mass spectrometry (GC–MS) is the primary tool for many of these methods. Examples of some of the specialist analysis methods used for fingerprinting are listed in Table 1. Unless noted otherwise, most methods are applicable to NAPLs, soil and groundwater.

Other types of analysis methods aim to attempt to determine when the contaminants were released to the subsurface or entered the groundwater. This is referred to as age dating. Examples of some of the specialist analysis methods used for quantitative age dating in subsurface forensics studies are shown in Table 2. Some of these methods aim to determine the date when the chemicals were produced whereas others aim to determine the date when the chemicals were released into the subsurface. Still others of these methods aim to determine the date when the water carrying the chemicals recharged to the groundwater and indirectly determine when the chemicals were released. Other methods may provide qualitative estimates of age. For example, over various periods different chemical additives such as alkylleads, ethylene dibromide (EDB), ethylene dichloride (EDC), methylcyclopentadienylmanganese tricarbonyl (MMT), methyl *tert.*-butyl ether (MTBE) and *tert.*-amyl methyl ether (TAME) have been used in gasolines.[14] The presence, absence or concentration level of such additive compounds in NAPL, soil or groundwater may provide minimum or maximum estimates of age depending on the history of use of the additive.

3.2.2.2 Modelling. Groundwater flow and contaminant migration modelling and vapour migration models are often part of routine hydrogeological studies. Analytical and numerical models of varying complexity are used in site characterization studies to help explain the present extent of groundwater contamination and to also make predictions of future migration and associated risks to receptors. Much modelling is undertaken to assist in the derivation of site-specific risk-based clean-up criteria (standards). Models may also be used to help design remedial measures such as groundwater pump-and-treat or *in situ* treatment. Many of these same models are applied in forensic studies, usually to attempt to answer specific questions, such as when did contamination commence and what were concentrations in the past? In some circumstances, very specialized models are used to assess, for example, the migration of contaminants through pavement (floors or sumps). Murphy and Morrison[4] provide a useful overview of the application of surface models designed to assess risks from subsurface contamination to surface-based receptors (as opposed to the underlying groundwater-based receptors).

Because of the nature of subsurface contaminant migration modelling, it is certain to be a target of severe scrutiny in forensic applications. Most models employ a relatively modest number of measured parameter values, combined with a substantial simplification of the actual subsurface conditions and many assumptions.

Groundwater contaminant migration models are used most often either to back-calculate the time when contamination commenced based on the measured extent of a groundwater plume, or to estimate the past contaminant concentrations or duration of contamination at a specific location such as a drinking water supply well. For both of these applications, some of the most

Table 1 Examples of specialist analytical methods used for fingerprinting in subsurface forensic studies.[a]

Specialist analytical method	Typical application
High-resolution GC–FID analysis of C_8–C_{44} hydrocarbons, evaluation of chromatographic patterns, quantification of specific *n*-alkanes and isoprenoids	Quantification of TPH Identification of petroleum products (*i.e.* fuels, other distillates, crude oil) Estimates of degree of weathering (*i.e.* evaporation, dissolution, biodegradation)
GC–MS analysis of C_5–C_{12} hydrocarbons, quantification of BTEX; fuel additives: oxygenates such as MTBE, TBA, TAME; other additives: EDB, EDC, MMT; PIANO group hydrocarbons: paraffins, isoparaffin, aromatics, naphthenes and olefins	Identification of light distillate products (*i.e.* gasolines, aviation fuels) BTEX concentrations and ratios, PIANO which may be characteristic of source Fuel additives which may be characteristic of source Estimates of degree of weathering (*i.e.* evaporation, dissolution, biodegradation)
High-resolution GC–MS analysis of C_8–C_{44} hydrocarbons, quantification of PAHs, alkyl-PAHs, *S*-, *N*-aromatics, biomarkers such as terpanes, steranes	Distinguish between petroleum-, coal- or combustion-derived hydrocarbons Assemblages of aromatics or biomarkers may be characteristic of source
GC–MS analysis of NAPL for alkyllead fuel additives, quantification of TEL, MTEL, DEDML, TMEL, TML	Assemblages of alkylleads may be characteristic of source
MS analysis of $^{15}N/^{14}N$ and $^{18}O/^{16}O$ stable isotope ratios in nitrate, $^{15}N/^{14}N$ in ammonia[38]	Isotopic composition may be characteristic of source Variations in isotopic composition may confirm geochemical processes
High-resolution GC–ECD or GC–MS analysis of PCBs, quantitation of individual congeners	Identification of specific Aroclor products or mixtures of Aroclors Estimates of degree of weathering (*i.e.* evaporation, dissolution, biodegradation)
GC–IRMS analysis of $^{13}C/^{12}C$ stable isotope ratios in specific organic compounds	Applied to various petroleum hydrocarbons, PAHs, oxygenates, chlorinated solvents $^{13}C/^{12}C$ ratio may be characteristic of source Variations in $^{13}C/^{12}C$ ratio may confirm biodegradation processes

[a]Abbreviations: GC–FID, gas chromatography–flame ionization detection; TPH, total petroleum hydrocarbons; GC–MS, gas chromatography–mass spectrometry; BTEX, benzene, toluene, ethylbenzene and xylenes; MTBE, methyl *tert.*-butyl ether; TBA, *tert.*-butyl ether; TAME, *tert.*-amyl methyl ether; EDB, ethylene dibromide; EDC, ethylene dichloride; MMT, methylcyclopentadienylmanganese tricarbonyl; TEL, tetraethyllead; MTEL, methyltriethyllead; DEDML, dimethyldiethyllead; TMEL, trimethylethyllead; TML, trimethyllead; MS, mass spectrometry; GC–ECD, gas chromatography–electron capture detection; GC–IRMS, gas chromatography–isotope ratio mass spectrometry.

Table 2 Examples of specialist analytical methods used for quantitative age dating in subsurface forensic studies.[a]

Specialist analytical method	Typical application
GC–MS analysis of C_5–C_{12} hydrocarbons, quantification of BTEX; PIANO group hydrocarbons: paraffins, isoparaffins, aromatics, naphthenes and olefins.	Age dating of gasoline[39] Toluene/n-octane ratio in gasoline has increased since 1970
High-resolution GC–MS analysis of C_8–C_{44} hydrocarbons, isoprenoids: pristane	Age dating of petroleum middle distillates (diesel, fuel oil) C_{17} n-alkane/pristane ratio declines with age due to differential weathering
TIMS analysis of $^{206}Pb/^{207}Pb$ isotope ratios	Age dating of manufacture of leaded gasoline additives Isotope ratio in lead used in alkylleads has varied systematically over time
GS–MS analysis of TCA and 11DCE in groundwater	Age dating of TCA releases from formation of 11DCE *via* abiotic degradation of TCA[40] 11DCE/TCA ratio increases with age
Dendroecology: microchemical analysis of tree rings	Past presence of subsurface contamination may be reflected by residual chemicals in specific rings
Radiometric analysis of tritium (^3H), stable isotope analysis of decay product ^3He in groundwater	Tritium originated from atmosphere nuclear weapons testing in 1953–1960s Absence of tritium indicates groundwater recharge before 1953. Tritium peak may indicate recharge in 1960s
GC–ECD analysis of CFC-11, CFC-12, CFC-113 in groundwater	CFCs in atmosphere since 1930s from anthropogenic releases Ratios have varied systematically over time

[a]Abbreviations: GC–MS, gas chromatography–mass spectrometry; BTEX, benzene, toluene, ethylbenzene and xylenes; TIMS, thermal ionization mass spectrometry; TCA, 1,1,1-trichloroethane; 11DCE, 1,1-dichloroethene; CFC-11, trichlorofluoromethane; CFC-12, dichlorodifluoromethane; CFC-113, trichlorotrifluoroethane.

important model parameters are those which are the least well known and must be estimated because they cannot be measured. These parameters include the historical concentrations and mass loading of contaminants emitted from the source zone, historical conditions in groundwater flow-rates and directions, dispersivity whether caused by geological heterogeneity or transient fluctuations in groundwater flow and the past and present rates of (bio)degradation.

More sophisticated modelling approaches continue to emerge[20] and become available to practitioners alongside the ever-increasing availability of greater computing power. Of relevance to subsurface forensics, for example, is the ongoing research effort directed to plume inversion – source reconstruction techniques whereby (stochastic) numeric models are used to reconstruct the historical source term from the observed groundwater plume data.[21] Applied uptake of such novel modelling tools, however, is typically slow for several reasons, including their increased cost or time requirements, model validation needs and acceptability in legal settings and the insufficiency of site data to populate models.

3.3   Historical Knowledge of Subsurface Contamination

Increasingly, forensic investigators are asked opinion on the historical knowledge regarding subsurface contamination by specific chemicals. Parties responsible for a contaminated site might be denied insurance coverage to pay for the investigation and clean-up or may be charged with punitive damages if the chemical releases or disposal took place at a time when the resultant contamination could have reasonably been expected. The authors are aware of litigation in which this been a question for groundwater contamination cases involving chlorinated solvents, perchlorate, PAHs, MTBE, dibromochloropropane (DBCP) and 1,2,3-trichloropropane (TCP).

Forensic work to develop such opinions generally relies on the review of information published during the relevant periods in the scientific, engineering and environmental regulatory fields and any information on the use, handling or disposal of the chemical products in question.

Not surprisingly, forensic experts on opposing sides of a case commonly express differing opinions despite relying on the same documentary record. Articles by Travis,[22] Jackson,[23,24] Amter and Ross[11] and Rivett et al.[25] give some flavour of the debate related to the historical knowledge of groundwater contamination by chlorinated solvents.

By way of example, the legal importance of foreseeability is illustrated by the UK's highest profile court case on chlorinated solvent pollution of a supply well.[26] The case was initially decided in favour of the polluter in the High Court, but later overturned in favour of the affected water company in the Court of Appeal, yet later overturned in favour of the polluter by the House of Lords. Common law principles applied and the House of Lords decided that the recovery of damages in nuisance depended on foreseeability and the polluter could not have reasonably foreseen the consequences of the escape of chlorinated solvent from its property at the time of release (judged to be prior to the mid-1970s).

4   Examples of Some Important and Challenging Specific Questions

The following sections outline examples of some important or challenging specific questions commonly posed in subsurface forensics cases. The discussion for each describes both the hydrogeological and forensic methods that may contribute to providing useful answers.

4.1   What Was the Chemical Material That Was Released?

Although this question might appear to be relatively simple, identification of the chemical material released may not be directly possible using routine sampling and testing of soil or groundwater because the materials have been altered by weathering processes (evaporation, dissolution, biodegradation). The following example relates to groundwater contamination by chlorinated solvents.

Tetrachloroethene, formerly known as perchloroethylene (PCE), and trichloroethene (TCE) are solvents commonly used in industry and have entered the subsurface from leaks, spills and disposal practices.[27] A given industrial site may have used just PCE, just TCE, both or switched from one to the other at some time. Knowing whether the material released at a site was PCE or TCE or both may be valuable because it may assist in determining either the activity or the site operator that may have been responsible for the release or the period when the release may have occurred. However, biodegradation of PCE via reductive dechlorination will form TCE. In turn, TCE degrades to *cis*-1,2-dichloroethene (c12DCE) and vinyl chloride (VC).[6] At many sites, PCE, TCE, c12DCE and VC are the important contaminants found together in various proportions in groundwater. The following sections describe how data from routine hydrogeological studies and/or specialist sample analysis methods might be used to answer this question.

4.1.1 Empirical Data. For example, the authors are aware of an industrial site (referred to here as XYZ), at which PCE, TCE and c12DCE concentrations in groundwater were found to be 56, 8100 and 920 $\mu g\,l^{-1}$, respectively. TCE is much higher than PCE, but the presence of c12DCE indicates that some degree of biodegradation has occurred. However, the absence of VC suggested that biodegradation may not have been extensive. At this site, the question arose of whether all the ethene series compounds were derived from a PCE release or whether there were also releases of TCE. Several lines of evidence can be considered to attempt to answer this question.

Rates of biodegradation of PCE to TCE and TCE to c12DCE are generally relatively rapid under suitable anaerobic conditions and the c12DCE to VC and VC to ethene steps are slower so that c12DCE or VC commonly accumulate in groundwater.[6] Although a forensic expert may testify that they have not previously observed accumulation of TCE in groundwater from reductive dechlorination of PCE, published or empirical information would be valuable to support such a conclusion.

For an example of such empirical data, the on-line database of the State Coalition for Remediation of Dry Cleaners (SCRDC) was reviewed. Maximum concentrations in groundwater of PCE, TCE, c12DCE and VC were reported for 132 sites of known PCE releases. Of these sites, data were examined in detail for 69 sites which had groundwater concentrations of $>1000\,\mu g\,l^{-1}$. Total ethene concentrations were calculated as the sum of the molar concentrations of PCE, TCE, c12DCE and VC expressed as $\mu g\,l^{-1}$ of PCE. Molar percentage concentrations were calculated for each compound. Examples of these calculations for several sites are shown in Table 3. PCE was the predominant compound at most of the sites (see Table 3, examples SCRDC A and B). At the relatively few sites at which TCE occurred at higher concentrations than PCE,

Table 3 Example of consideration of empirical data to assess whether releases were PCE or TCE. Proportions of perchloroethene (PCE), trichloroethene (TCE), *cis*-1,2-dichloroethene (c12DCE) and vinyl chloride (VC) from known PCE releases are compared with those at a site where the nature of the release was not known.[a]

Site	Total ethenes $(\mu g\,l^{-1})$	PCE (%)	TCE (%)	c12DCE (%)	VC (%)	Nature of release
SCRDC A little biodegradation	39699	99	0.16	0.77	–	Known PCE
SCRDC B moderate biodegradation	30686	58	3.4	38	0.05	Known PCE
SCRDC C extensive biodegradation	36860	5.4	40	47	7.3	Known PCE
SCRDC lowest PCE/TCE ratio	10154	3.2	11	70	12	Known PCE
Site XYZ with unknown release	11890	0.5	86	13	–	Probable TCE

[a]Empirical data are the maximum groundwater concentrations from the web database 12 June 2007 of the State Coalition for Remediation of Dry Cleaners (SCRDC) in the USA. Total ethene concentrations expressed in $\mu g\,l^{-1}$ as PCE. PCE, TCE, c12DCE and VC concentrations expressed as molar percentages of total ethene concentration.

c12DCE concentrations were comparable to or higher than those of TCE, indicating that a high degree of biodegradation had occurred (examples SCRDC C and D). Based on comparison with this empirical information, it would be unlikely that the groundwater contamination found at site XYZ resulted from the release of only PCE.

4.1.2 Summed Molar Concentrations Versus Solubility. In some circumstances, the concentrations of degradation products in the groundwater may be sufficiently high to eliminate the possibility that a release was principally PCE. For example, at a US Air Force Base site,[6] groundwater concentrations of PCE, TCE, c12DCE and VC were 56, 15 800, 98 500 and 3080 $\mu g\,l^{-1}$, respectively. If all of these compounds derived from a PCE source, their sum would exceed the solubility of PCE. Given factors such as the spatial variability of DNAPL source zones and dilution/dispersion and mixing as a result of groundwater sampling, the sum of these degradation products would not commonly exceed 10% of the PCE solubility.[28] However, in this example the sum of the molar concentrations of the ethene compounds, expressed as PCE, is 197 000 $\mu g\,l^{-1}$, and this value exceeds the common literature value of 150 000 $\mu g\,l^{-1}$ for the solubility of PCE. In this case it would be concluded that the principal source of the groundwater contamination was a release of TCE rather than a release of PCE.

4.1.3 Stable Isotopic Methods. The examination of $^{13}C/^{12}C$ isotope ratios has the potential of determining whether TCE is derived from TCE releases or from biodegradation of PCE. $^{13}C/^{12}C$ ratios in specific organic compounds such as PCE and TCE can be measured using gas chromatography–isotope ratio mass spectrometry (GC–IRMS). Isotopic composition is not expressed directly as a ratio but as per mil (‰ or parts per thousand) deviation of the ratio in comparison with a recognized standard reference material. An example of the conventional form to report the stable carbon isotopic composition of PCE in a sample is $\delta^{13}C = -30‰$. This means that the $^{13}C/^{12}C$ ratio in the PCE is 30‰, 30 per 1000, or 3% lower than the ratio in the standard reference material. An overview of the use of stable isotopes in groundwater studies was given by Slater.[29]

The examination of stable carbon isotope compositions in PCE and other chlorinated solvents has been used to provide insight into biodegradation processes in the subsurface environment because fractionation of the isotopes occurs during biodegradation. Because ^{13}C has a slightly higher atomic weight, chemical bonds in molecules of PCE containing a ^{13}C atom are slightly stronger and these molecules biodegrade at a slightly slower rate than molecules containing only ^{12}C atoms. Consequently, as PCE degrades to TCE, the TCE formed by the reaction is depleted in ^{13}C and the $\delta^{13}C$ of the TCE is lower than the PCE from which it was formed. The PCE remaining during the reaction becomes enriched in ^{13}C and its $\delta^{13}C$ increases progressively. The isotopic changes are comparable for the degradation of TCE to c12DCE.

In the PCE to TCE to c12DCE degradation sequence, if all the TCE and c12DCE were derived from PCE, the $\delta^{13}C$ of the TCE would be always lower than that of the PCE and the $\delta^{13}C$ of the c12DCE would be always lower than that of the TCE. In laboratory microcosm studies of the biodegradation of PCE,[30,31] this pattern is observed although differences in $\delta^{13}C$ between TCE and c12DCE were substantially greater (2–15‰) than the differences in $\delta^{13}C$ between PCE and TCE ($< 2‰$). If the $\delta^{13}C$ of each degradation product in the PCE to TCE to c12DCE sequence were progressively lower (more negative), this pattern would be evidence that TCE and c12DCE were both derived from degradation of the PCE. However, if this pattern is not observed, it could be the result of variations in the $\delta^{13}C$ of the PCE in the source(s), differences in relative rates of the degradation steps or because there were also releases of TCE.[32,33]

Table 4 shows an example of $\delta^{13}C$ analyses of PCE, TCE and c12DCE in groundwater at an industrial site that repackaged and distributed various chemical products over a period of several decades including PCE and perhaps TCE. In the area nearest the suspected source zones concentrations were the highest, the groundwater contained PCE, TCE and c12DCE indicate biodegradation had occurred. Questions posed for this case included: Did the TCE originate solely from biodegradation of PCE or was there also release of TCE; Did the PCE originate from one release or multiple releases?

The groundwater sample from location C had concentrations of c12DCE much higher than those of PCE and TCE, indicating substantial biodegradation with accumulation of c12DCE. Vinyl chloride concentrations were commonly below the detection limit or no higher than about $1.5\,\mu mol\,l^{-1}$. Groundwater at locations A and D had concentrations of PCE higher than those of TCE and c12DCE, indicating a lesser degree of biodegradation at these locations.

Table 4   Example of $\delta^{13}C$ analyses of groundwater at an industrial site.[a]

Location	PCE ($\mu mol\,l^{-1}$)	TCE ($\mu mol\,l^{-1}$)	c12DCE ($\mu mol\,l^{-1}$)	$\delta^{13}C$ PCE (‰)	$\delta^{13}C$ TCE (‰)	$\delta^{13}C$ c12DCE (‰)
A	91	34	22	−32.0	−34.2	−32.8
B	130	66	12	−30.4	−29.8	−30.2
C	4.5	1.8	440	−29.9	−25.3	−32.3
D	120	50	65	−26.7	−28.1	−27.4

[a]Abbreviations: PCE, perchloroethene; TCE, trichloroethene; c12DCE, *cis*-1,2-dichloroethene; $\delta^{13}C$ expressed as ‰ deviations from V-PDB reference standard.

Groundwater at location B had the highest concentration of PCE and the lowest concentration of c12DCE, indicating the lowest degree of biodegradation.

The groundwater at locations A and D exhibited similar concentrations and a similar degree of biodegradation; however, the $\delta^{13}C$ of PCE, TCE and c12DCE at location A are all lower (more negative) by 5–6‰ than at location D. The fact that the $\delta^{13}C$ of PCE at the two locations is substantially different while exhibiting a similar degree of biodegradation suggests that the PCE at these two locations originated from different releases of PCE having different initial isotopic compositions. These $\delta^{13}C$ values for PCE fall within the range reported for PCE product (approximately −35 to −25‰) from different manufacturers.[34] The reported range of $\delta^{13}C$ values for TCE product from different manufacturers is similar to that for PCE.

The patterns of $\delta^{13}C$ values at each location for the sequence PCE–TCE–c12DCE do not clearly show progressively lower values, which would be expected if all the compounds originated from PCE and effect of biodegradation was the principal control on the $\delta^{13}C$ values. Deviations from the expected pattern could be the result of contribution of TCE with different isotopic compositions that originated from TCE releases. These results leave open the possibility that there were releases of TCE at the site.

4.2   Does the Groundwater Plume Track Back to the Releases or Source Zones?

When contamination of groundwater is identified, hydrogeological investigations are normally performed to determine the locations of the source or sources of the contamination. Such investigations usually involve interpretation of groundwater flow directions followed by installation of borings and/or monitoring wells upgradient of the known groundwater contamination to attempt to 'track' the plume back to the source zone(s) or release area(s). This approach is appropriate when the groundwater plume is 'attached' to a persistent source caused by ongoing releases, the presence of leachable solid wastes or by the presence of NAPL in the subsurface. The highest concentrations of contaminants in the groundwater are usually found close to such sources.

As illustrated in Figure 1, if releases to the subsurface were in the form of aqueous discharges (*i.e.* wastewater or leachate) only and either ceased or diminished substantially in times past, the groundwater plume may be 'detached' from the release areas and would appear as a pulse plume some distance downgradient of the release area. Similar detached plumes may arise also from highly soluble components of a NAPL source which have been rapidly depleted. MTBE from a gasoline source is an example of a highly soluble component that may form a detached plume in some circumstances. In general, such detached groundwater plumes are relatively rare. The most common organic contaminants in groundwater, petroleum hydrocarbons and chlorinated solvents are derived almost exclusively from LNAPL or DNAPL sources that have persisted in the subsurface for many years or decades after their release to the subsurface.

Although it is taken almost for granted by most hydrogeologists that groundwater plumes of petroleum hydrocarbons or chlorinated solvents track back to their source zone, even such a simple issue may be argued in litigation. The authors are aware of a case in which one of the questions at issue was whether the TCE plume represented a detached plume derived from a source far upgradient and on another site or whether the source of the plume was located near to the areas of highest groundwater concentrations. The TCE plume in the groundwater had a localized zone of relatively high concentrations (about $10\,000$–$20\,000\,\mu g\,l^{-1}$ in an area of about $25 \times 25\,m$). TCE concentrations declined progressively in a downgradient direction over a distance of at least $100\,m$ and declined markedly in lateral and upgradient directions. The notable degree of dispersion the downgradient direction and lack of 'tailing' in concentrations in the upgradient direction would argue against a detached plume.

Closely spaced soil borings and soil sampling had not yet been completed in the area of highest TCE concentrations to determine directly if the vadose zone or groundwater zone in this area contained 'the source' of the contamination. For TCE plumes, such a source would most commonly be an appreciable quantity of TCE mass in the form of residual or free-phase NAPL or alternatively zones in which TCE previously in NAPL form has dissolved and migrated by molecular diffusion into finer-grained strata (*i.e.* low-permeability clay layers or porous rock matrix adjacent to fissures).[35] Sufficiently detailed soil sampling and analysis could confirm the presence of such a source. However, the magnitude of the TCE concentrations in groundwater were sufficiently high (1–2% of pure-phase solubility) to indicate the presence of a source zone in the area of the highest groundwater concentrations.[6,36] In such a case, the forensic expert must weigh the confidence in the interpretation of an attached plume and identification of the general area of the source zone based primarily on the groundwater concentrations and configuration of the plume, *versus* insisting on the need for further soil borings and sampling.

4.3 Can the Contaminants be Traced Back to the Source Zone?

Depending on the circumstances of any particular contaminated site, there may be a variety of specialist fingerprinting analysis methods that can help trace contaminants back to their source based on their chemical characteristics. Many more methods are available for petroleum hydrocarbons rather than chlorinated solvents (and other contaminants of concern), because petroleum fuels and hydrocarbon products are usually comprised of many more compounds and more possibilities for comparing and contrasting chemical compositions.

Sometimes, the application and interpretation of such methods are subject to no complexities. Table 5 shows an example of the use of MTBE/TAME ratios to distinguish between two potential sources of MTBE contamination of a regional water supply aquifer. The MTBE/TAME ratios in the regional aquifer were high (average 2800) and were consistent in the contaminated monitoring wells. Only the groundwater in the source area of site A had MTBE/TAME ratios comparable to those found in the regional aquifer. MTBE/TAME ratios in the source area of site B and in some areas of site A were far too low to account for the contamination found in the regional aquifer.

The application and interpretation of fingerprinting methods is more complex at old sites (industrial, commercial or retail) that have used chemical products over long periods. Different chemical products may have been

Table 5 Example of using methyl *tert.*-butyl ether (MTBE)/*tert.*-amyl methyl ether (TAME) ratios in groundwater to distinguish different gasoline sources. Data from Feenstra.[41]

Location	MTBE $(\mu g\,l^{-1})$	TAME $(\mu g\,l^{-1})$	MTBE/TAME ratio
Regional aquifer MW-1	14000	4.0	3500
Regional aquifer MW-2	4400	2.0	2200
Regional aquifer MW-3	3000	1.0	3000
Regional aquifer MW-4	1600	0.68	2400
Average			**2800**
Site A MW-1	62000	36	1700
Site A MW-2	22000	11	2000
Site A MW-3	14000	5.0	2800
Site A MW-4	12000	3.0	4000
Site A MW-5	9500	7.2	1300
Site A MW-6	5100	2.3	2200
Site A MW-7	590	27	22
Site A MW-8	74	3.0	25
Site A MW-9	28	2.0	14
Site A MW-10	27	2.0	14
Average			**1400**
Site B MW-1	2500	43	58
Site B MW-2	1400	17	82
Site B MW-3	1200	16	75
Site B MW-4	1000	15	67
Site B MW-5	870	12	73
Site B MW-6	750	2.0	380
Site B MW-7	670	13	52
Site B MW-8	480	10	48
Site B MW-9	300	4.0	75
Site B MW-10	160	4.0	40
Site B MW-11	130	2.0	65
Average			**91**

released at different times. The chemical composition of a specific product may have changed over time. There may have been multiple releases over time of different chemical materials at the same locations. Weathering processes (evaporation, dissolution, biodegradation) have altered the chemical composition of the chemical materials after their release to the subsurface.

Because of these factors, the chemical characteristics of the original releases may not be known with much certainty. Modern fingerprinting methods offer sophisticated and rigorous ways to compare and contrast the chemical characteristics of NAPL, soil or groundwater from present-day investigations;[37] however, forensic opinions regarding the question WHO DONE IT? may be often limited because the same chemical characteristics of the original releases are not known precisely.

The following is an example of how these unknowns may be manifest in determining the origin of a gasoline source. In this case there were two retail gasoline stations (sites A and B) and gasoline product was discovered in a monitoring well on a neighbouring property (neighbour site). Gasoline was found also in several monitoring wells at both sites A and B. Gasoline may have migrated to the neighbour site from site A directly, from site B directly or from site A through site B. The sites are close together, the geological strata are relatively complex and the flow directions may have been influenced by leaking water mains and sewers, and hence there is uncertainty in groundwater flow paths. Site A operated from 1942 until 1985 and site B from 1954 until 2007. The issue of the case was to determine whether site A or site B was responsible for the gasoline on the neighbour site.

High-resolution GC–MS analyses were performed on LNAPL samples from two monitoring wells on site A, one well on site B and the one-well neighbour site. Concentrations of PIANO (paraffins, isoparaffins, aromatics, naphthenes and olefins) components, alkylleads and MMT (methylcyclopentadienylmanganese tricarbonyl) were determined and evaluated together with total lead, total manganese and sulfur. Table 6 summarizes some of the aspects of this evaluation.

In this case, neither of the parties representing both site A and site B was able to pass all responsibility to the other. The gasoline on the neighbour site did not resemble closely any of the other samples. The samples showed different degrees of weathering, with the gasoline on the neighbour site being the least weathered. The concentrations of tetraethyllead (TEL) indicated that the samples were wholly or predominantly leaded fuel. In this jurisdiction this indicates fuel from before about 1987. One of the samples had MMT indicating post-1990 fuel, but at the same time had TEL indicating pre-1987 fuel, suggesting the gasoline at site B was a mixture of older and more recent releases. The isooctane ratio in the sample from site B indicated alkylated gasoline whereas all other samples were reformate gasolines. Other alkyllead compounds were present on site A but absent on site B and the neighbour site. Based on this fingerprinting exercise, it was not possible to determine the source of the gasoline found on the neighbour site because there have probably been multiple releases over time at both sites A and B. Some mixing of gasoline on each site and between sites may have occurred to blend any chemical fingerprints.

Table 6   Example of gasoline fingerprinting at old (>50 years) retail gasoline
stations.[a]

Location	Degree of weathering	TEL (mg kg^{-1})	Other alkyl-leads (mg kg^{-1})	MMT (mg kg^{-1})	Isooctane ratio
Site A MW-1	Moderate	490[b]	130	<1	1.9
Site A MW-2	Low	800	48	<1	1.0
Site B MW-1	Moderate	380	<1	95[c]	10.0[d]
Neighbour site	Very low	730	<1	1.0	0.8

[a]Abbreviations: TEL, tetraethyllead; MMT, methylcyclopentadienylmanganese tricarbonyl;
  isooctane ratio, isooctane/methylcyclohexane.
[b]Indicates pre-1987 gasoline.
[c]Indicates post-1990 gasoline.
[d]Indicates alkylated gasoline. Other results indicate reformate gasolines.

Even without mixing, the compositions of the petroleum products from
specific refiners at specific times from decades ago are known only approximately. No fingerprint analyses can be available for the various gasoline
products that may have been released at the sites far in the past, in order to
compare with samples and analyses from the present.

In addition to these factors, weathering processes are dependent on the
gasoline volumes, evaporative losses, rates of rainfall infiltration and groundwater flow through the source zone. These processes are not uniform in the
subsurface and thus the degree of weathering of the gasoline may not be uniform even for releases of the same product at the same time.

Considerable forensic expertise and experience in fingerprinting methods are
necessary, in combination with fundamental hydrogeological information, to
develop sound, legally robust conclusions regarding contaminant sources at
many contaminated sites.

5   Concluding Discussion

Past practices employed for the use, handling, storage and disposal of various
chemical products have caused groundwater pollution at a great many sites
throughout the developed world. Because of the damage to private and public
property arising from such contamination and because of the high costs associated with investigation and remediation of such contamination, an increasing number of such sites will become the subject of litigation to identify the
parties responsible for such costs. As a result, the need for subsurface forensics
studies will increase in the foreseeable future to help answer the questions
Who?, Where? and How? Many specialist sample analysis methods have been

developed in recent years and applied to help answer these questions. There is little doubt that historical groundwater pollution represents a significant environmental forensic challenge. For the further growth of subsurface forensics, it is essential that specialist forensic sample analysis and modelling methods be increasingly developed and diversified and combined closely with the various elements of more routine contaminant hydrogeological studies, assessment of site history and knowledge of chemical uses and practices.

References

1. R. D. Morrison and B. L. Murphy, *Environmental Forensics: Contaminant Specific Guide*, Academic Press, San Diego, CA, 2005.
2. R. D. Morrison, *Environmental Forensics: Principles and Applications*, CRC Press LLC, Boca Raton, FL, 2000.
3. P. J. Sullivan, F. J. Agardy and R. K. Traub, *Practical Environmental Forensics: Process and Case Histories*, Wiley, New York, 2001.
4. B. L. Murphy and R. D. Morrison, *Introduction to Environmental Forensics*, 1st edn., Academic Press, San Diego, CA, 2002.
5. B. L. Murphy and R. D. Morrison, *Introduction to Environmental Forensics*, 2nd edn., Academic Press, San Diego, CA, 2007.
6. T. H. Weidemeier, H. S. Rafai, C. J. Newell and J. T. Wilson, in *Natural Attenuation of Fuels and Chlorinated Solvents in the Subsurface*, Wiley, New York, 1999.
7. R. E. Jackson, *Hydrogeol. J.*, 1998, **6**, 144.
8. M. O. Rivett and S. Feenstra, *Environ. Sci. Technol.*, 2005, **39**, 447.
9. C. Eberhardt and P. Grathwohl, *J. Contam. Hydrol.*, 2002, **59**, 45.
10. D. M. Mackay and J.A. Cherry, *Environ. Sci. Technol.*, 1989, **23**, 630.
11. S. Amter and B. Ross, *Environ. Forensics*, 2001, **2**, 179.
12. R. Doherty, *Environ. Forensics*, 2000, **1**, 69.
13. R. Doherty, *Environ. Forensics*, 2000, **1**, 83.
14. L. M. Gibbs, *Gasoline Additives – When and Why*, SAE Technical Paper Series 902104, Society of Automotive Engineers, Warrendale, PA, 1990.
15. J. Lohman, *Environ. Forensics*, 2002, **3**, 35.
16. M. Hamper, *Enironv. Forensics*, 2006, **7**, 55.
17. US Environmental Protection Agency, *Test Methods for Evaluating Solid Waste SW-846 On-Line*, EPA, Washington, DC, 2007.
18. American Public Health Association, *Standard Methods for the Examination of Water and Wastewater On-Line*, APHA, Washington, DC, 2007.
19. J. F. Pankow, S. Feenstra, J. A. Cherry and M. C. Ryan, in *Dense Chlorinated Solvents and Other DNAPLs in Groundwater*, ed. J. F. Pankow and J. A. Cherry, Waterloo Press, Beaverton, OR, 1996, p. 1.
20. J. Atmadja and A. C. Bagtzoglou, *Environ. Forensics*, 2001, **2**, 205.
21. A. Woodbury, E. A. Sudicky, T. J. Ulrych and R. Ludwig, *J. Contam. Hydrol.*, 1998, **32**, 131.
22. A. S. Travis, *Mealey's Emerging Toxic Torts*, 1998, **7**, 41.
23. R. E. Jackson, *Environ. Eng. Geosci.*, 1999, **5**, 331.

24. R. E. Jackson, *Environ. Forensics*, 2003, **4**, 3.
25. M. O. Rivett, S. Feenstra and L. Clark, *Environ. Forensics*, 2006, **7**, 313.
26. R. P. Ashley, in *Groundwater Contaminants and Their Migration*, ed. J. D. Mather, Special Publication 128, Geological Society, London, 1998, p. 23.
27. M.O. Rivett, K.A. Shepherd, L. Keeys and A.E. Brennan, *Q. J. Eng. Geol. Hydrogeol.*, 2005, **38**, 337.
28. S. Feenstra and J. A. Cherry, in *Dense Chlorinated Solvents and Other DNAPLs in Groundwater*, ed. J. F. Pankow and J. A. Cherry, Waterloo Press, Beaverton, OR, 1996, p. 395.
29. G. F. Slater, *Environ. Forensics*, 2003, **4**, 13.
30. D. Hunkeler, R. Aravena and B. J. Butler, *Environ. Sci. Technol.*, 1999, **33**, 2733.
31. G. F. Slater, B. Sherwood Lollar, B. E. Sleep and E. A. Edwards, *Environ. Sci. Technol.*, 2001, **35**, 901.
32. B. M. Van Breukelen, D. Hunkeler and F. Volkering, *Environ. Sci. Technol.*, 2005, **39**, 4189.
33. A. Vieth, J. Muller, G. Strauch, M. Kastner, M. Gehre, R. U. Meckenstock and H. H. Richnow, *Isotopes Environ. Health Stud.*, 2003, **39**, 113.
34. N. Jendzejewski, H. G. M. Eggenkamp and M. L. Coleman, *Appl. Geochem.*, 2001, **16**, 1021.
35. S. Feenstra, J. A. Cherry and B. L. Parker, in *Dense Chlorinated Solvents and Other DNAPLs in Groundwater*, ed. J. F. Pankow and J. A. Cherry, Waterloo Press, Beaverton, OR, 1996, p. 53.
36. C. J. Newell, R. R. Ross, *Estimating Potential for Occurrence of DNAPL at Superfund Sites*, US Environmental Protection Agency Publication 9355.4-07FS, EPA, Washington, DC, 1991.
37. Z. Wang and S. A. Stout, *Oil Spill Environmental Forensics: Fingerprinting and Source Identification*, Academic Press, Burlington, MA, 2007.
38. I. Clark and P. Fritz, *Environmental Isotopes in Hydrogeology*, Lewis, Boca Ratio, FL, 1997.
39. G. Schmidt, *Environ. Forensics*, 2002, **3**, 145.
40. T. Gauthier and B. Murphy, *Environ. Forensics*, 2003, **4**, 205.
41. S. Feenstra, *Environ. Forensics*, 2006, **7**, 175.

Subject Index